Writing & Designing

OPERATOR MANUALS

Including Service Manuals and Manuals for International Markets

Gretchen H. Schoff • Patricia A. Robinson

LIFETIME LEARNING PUBLICATIONS
Belmont, California

A division of Wadsworth, Inc.

London, Singapore, Sydney, Tokyo, Toronto, Mexico City

T
11
.S376
1984

Designer: Richard Kharibian
Production Editor: Richard Mason
Copy Editor: Steve Bodian
Composition: Thompson Type

Printed in the United States of America

1 2 3 4 5 6 7 8 9 0 — 87 86 85 84

Library of Congress Cataloging in Publication Data
Schoff, Gretchen H.
  Writing and designing operator manuals.

  Bibliography: p.
  Includes index.
  1. Technical writing.   I. Robinson, Patricia A.
  II. Title.
T11.S376   1984   808'.0666   84-3865
ISBN 0-534-03362-8

Contents

List of Figures

Preface

Operator manuals give verbal and visual instructions for the use of thousands of products, ranging from toasters and tractors to cameras and cash registers. Almost every product except the simplest comes with such instructions. These instructions are known as operator manuals, owner manuals, user manuals, or simply instructions. They are "how to" books for owners and operators of the products.

This book also is a "how to" book—written for those who produce manuals. Our purpose is to show them the most effective ways to create manuals for products. In creating this book we have been guided by our experience in teaching and consulting with writers and designers from many different kinds of businesses and industries.

The Audience for This Book

Writers, artists, engineers, product safety personnel, photographers, editors, technicians, and managers make up the audience for this book. The book should be useful to writers who want to improve their company manuals, who are assigned to produce a manual for the first time, or who are creating manuals for newly developed products. It should also be useful to producers of technical drawings and photos and to those responsible for safety programs and product liability matters.

Talking with manual writers has helped us to shape the book effectively. We have found that many manual writers were not trained as professional technical writers, but rather entered the field of manual writing "sideways," through another profession or another unit of their company. On the other hand those with training in writing often feel they lack expertise in engineering concepts, visuals, or graphics; most of the writers we have met have learned on the job—and were often left to their own devices to figure out what needed to be done. This book offers guidelines for all the facets of manual production, from writing clear instructions to choosing the right binding, so as to enable both the professional without a technical writing background and the technical writer without an engineering background to write effective manuals for a diversity of products.

The Importance of Manuals as Publications

At first glance, operator manuals and service manuals may seem to involve little more than instructions for use and care of the product. In reality, manuals do much more.

Instructions

Most operator manuals contain instructions for assembly, operation, maintenance, and storage of products. Many manuals also contain sections on troubleshooting, service, and repair. Very complex products, such as electronics, heavy industrial machinery, or biomedical devices, usually have separate manuals for service and repair. New products, for which no prototype manuals exist, and complex products, in which many subsystems interact, present especially difficult problems for writers. Such manuals demand that the writers have a grasp of an astonishing diversity of mechanisms, processes, and procedures. Therefore, writers must continually be asking themselves, "How can I best describe or show how this works?"

Product Liability Document

The consumer protection movement and the legal climate surrounding product liability law are of serious concern to manufacturers of any product. Of special current interest are the safety warnings found throughout operator manuals and the safety labels actually affixed to the product. One important component of product liability law is the manufacturer's "duty to warn." Manuals must warn product users against such hazards as electrical fields, sharp blades, moving parts, shattering glass, chemicals, toxic substances, and flammable and explosive materials. The operator manual and its warnings frequently become key documents in product liability suits. Therefore, if the manual is well designed and worded, it may help protect the manufacturer against charges of failure to give adequate warning.

An Advertisement for the Product and a Part of the Company Image

The operator manual sends a message to the buyer of your product. A poorly designed, confusing, or unreadable manual may cast doubt on the quality of the product itself or convince a user not to buy from a certain company in the future, even if the products of that company are good. Think, for instance, of a buyer who tries out his new camera for the first time and finds the manual so hard to use that he can't put the film in without returning to the photo shop. His teeth are set on edge even before he has had a chance to try the product. In contrast, top-quality manuals are read, used, and saved by the owner—and often envied and copied by competitors.

Major Sections of the Book

We have organized the book along the same lines as the major steps taken in writing the manual: planning; analyzing the user; choosing organizational and writing strategies; coordinating format, references, and mechanics; and creating visuals. We also include chapters on the following special topics: safety messages, service manuals, manuals for international markets, and manual production. Writers make many choices as they put together a manual. We describe the decisions that must be made and show which techniques work best. Careful control of the manual elements produces a document that is clear, technically correct, and readable by the widest possible range of users. Throughout our book—but especially in Chapter 1, "Planning," and Chapter 9, "Managing and Supervising Manual Production,"—we discuss techniques for organizing office structures and writing teams to make the manual production process run as smoothly as possible.

Approach and Features of the Book

We have tried to keep the language of this book clear, direct, and unacademic. The principles it discusses can be applied in writing manuals for a diversity of products (household, automotive, chemical, industrial, mechanical, recreational, biomedical). Throughout the book, you will find many examples and illustrations. For instance, the chapter on analysis of user needs provides a series of questions that can be used to make such an analysis. The chapter on visuals provides samples of effective photos and drawings, plus advice on how to produce them. The chapter on safety warnings shows how to design them and how to avoid ambiguous wording.

If, as a writer on the job, you are faced with a deadline, and you are writing copy, choosing paper and type size, hiring a printer, laying out pages, producing drawings and photos (or contracting for someone to do them)—this book will help you come up with a manual that helps to protect both the user and the manufacturer of your product.

Gretchen H. Schoff
Patricia A. Robinson

Acknowledgments

Many individuals have contributed to this book by sharing their expertise and their materials or by allowing us to observe and work with their writers. We owe a special thank you to the following people: John Conrads, General Services Manager, Deere and Company; John Gormley, Director, Product Safety, Westinghouse Electric Corporation; Thomas Hillstrom, Product Safety Division, International Harvester Company; Albert O. Hughes, Supervisor, Special Operations, FMC Corporation; Fred Lineberry, Technical Publications, International Harvester Company; W. Robert Marshall, Director, University-Industry Research, University of Wisconsin; Richard Moll, Professor, Engineering and Applied Science, University of Wisconsin–Extension; Charles Robert Pearsall, Manager, Service Publications, Deere and Company; Fred Rode, Manager, Technical Training Literature, Outboard Marine Corporation; Delmar Swann, Personnel Development Division, E. I. DuPont De Nemours and Company; John Thauberger, Senior Engineer, Prairie Agricultural Machinery Institute; Ivan Thue, General Manager, Prairie Implement Manufacturers' Association.

The materials for the book were collected in a number of ways. We have conducted in-house technical writing seminars for industries, visited service publications operations, lectured and taught in national and regional conferences and workshops devoted to product safety and technical writing, and served as private consultants. Individuals and organizations who have worked with us and/or whose materials provided the subjects and examples for this book include: Acme Burgess Inc; American Optical Company; Atwood Mobile Products; Ann Bitter; Butler Manufacturing Company; Chrysler Motors Corporation, Dodge Division; Construction Industry Manufacturers' Association; CooperVision Surgical Systems; Deere and Company; Devilbiss Company; Doboy Packaging Machinery; E. I. DuPont De Nemours and Company; Farm Implement and Equipment Institute; FMC Corporation; Ford Motor Company, Ford Tractor Division; Fulton Manufacturing Corporation; General Electric Company; General Motors Company; Gerber Products Company; Harley Davidson Motor Company; Hazelton-Raltech Incorporated; Honeywell Inc.; Huffy Corporation; Ingersoll-Rand Company; International Business Machines; International Harvester Company; J. I. Case Company; John Muir Publications; Kohler Corporation; James J. McNeary; Madison-Kipp Corporation; Norden Laboratories; Ohio Medical Products; Outboard Marine Corporation; Robert Perras; Prairie Agricultural Machinery Institute; Prairie Implement Manufacturers' Association (and affiliated companies of Canada); Rosemount Inc.; Siemens Medical Systems; Silver-Reed America, Inc.; Teresa Sprecher; Rosemary Stachel; Sunstrand Aviation; Taylor Instrument Company; Technicare; Thern Inc.; Versatile Corporation; Volkswagen of America; Wabco Construction and Mining Equipment; Westinghouse Electric Corporation; and Yamaha Motor Company.

1

Planning

Overview

In our preface, we note that manual writers from many businesses have been our guides in creating this book. They have asked us questions that led us to look for answers, and they have generously shared with us their favorite tricks of the trade. We have also found that whether writers are doing manuals for copiers or paint sprayers, for cash registers or trailer hitches, they encounter similar problems. Writers often see those problems as the direct result of poor planning before actual manual production starts.

This chapter treats the planning process. Whether manuals are produced by a single writer or by a team, careful planning does much to set the stage for effective manual production. We begin by looking at the planning necessary for technical writers who work singly or in teams. Then we look at preliminary planning for the manual itself. Because writers need information and time, we give suggestions on how to fulfill these basic needs. Many of these suggested techniques can be used not only by writers, but by publications managers and supervisors as well. (See Chapter 9, "Managing and Supervising Manual Production," for fuller treatment of supervisory techniques.)

The Solo Writer

Small companies often assign one person to do the manuals for products. If you are that solo writer, you will soon find that writing is only one part of the job. You may also have to do photography, plan the art work, choose paper stock, edit, and type.

Advantages

As a solo writer, you have many opportunities to be creative. Because the majority of decisions will fall to you alone, you can approach the manual production job with your own vision of how the final manual will look, and you can make certain decisions without having to clear each step of the production with someone else. We have met a number of solo writers who say the autonomy they enjoy more than compensates for their many responsibilities. They also value the variety of tasks involved and enjoy the different kinds of people they work with. Most of all, they like having control over the project from start to finish.

Disadvantages

If you are a solo writer, your work will be the single bridge between the technical data about your product and the manual that reaches users. You will have to gather the information and create the schedule yourself. Manual writers who work solo often feel rushed, isolated, and pressured by their many responsibilities. They sometimes feel that other personnel, those on whom they must rely for information, have scant understanding of what it takes to put a manual together.

Making Solo Writing Easier

Your needs as a solo writer are much the same as the needs of team writers. You need access to information and time to do the job. As you read the rest of the chapter on team writing and on information and time, you will find many suggestions that you can adapt to the solo-writing setting.

Consider, especially, ways in which you can perform the same functions as a team leader performs in team writing. For example, you can do your own advance planning by

- Setting up a style and format handbook or a set of guidelines so that your own writing procedures become standardized and easier to repeat from manual to manual
- Arranging your own schedule of meetings with key personnel to collect information
- Asking for help from informal support teams or individuals (for work such as typing, drawing, planning safety messages, taking photographs)
- Developing a thorough outline
- Laying out steps in manual production

Team Writing

Manual writing is often done by a team of writers, especially if the product is complex. Such division of labor makes sense for a number of reasons: preparation time can be shortened, writers can develop special expertise with certain manual segments, and teams can include personnel from other company units (e.g., technical, research, product safety). The team-written manual also poses problems, particularly those of conceptual unity, team coordination, and uniformity of quality. Here are some of the pros and cons of the team-written manual and some suggestions for making team writing efforts smoother.

Advantages

"Many hands make light work" goes the saying. Dividing the manual writing according to systems or processes inherent in the product or according to special areas of writer expertise allows you to make the best use of writer talent and to get the job done more quickly and accurately. For example, the writer whose specialty is filtration systems, calibration, or electrical systems will find it easier to write about that area than the writer who has to keep many different kinds of processes or procedures in mind. Situations also arise in which a machine or product has used standard mechanical or chemical processes and is then suddenly altered by new technology or by the addition of an electronic component. In such cases, the best use of talent may be to ask the technician or engineer-designer who created the new component to write the segment describing its function. The most frequent kinds of product alterations in the last decade have been those involving computers, numerical control, or robotics (e.g., devices for welding, spraying, assembly procedures, and quality control).

Disadvantages

‘‘The camel is an animal designed by a committee’’ goes another saying. Too often, the team-written manual has camel-like lumps and bumps. Such manuals move by fits and starts from one segment to another. They sometimes have ill-matched writing styles and formats. Users find them very hard to use because of their redundancy, their lack of cross-referencing, and their chaotic organization. In brief, the chief difficulty with the team-written manual is the coordination of several writers' work into a smooth manual that looks as if one person had written it.

Coordinating the Team Effort

Team-written manuals must have a team leader, a manager, or a service publications editor who has the final responsibility for and a unified concept of what the finished manual will look like. That unified concept may be the joint creation of the writing team; however, once the conceptual framework is established, one person should be responsible for the scheduling, the assignment of manual segments, the creation of clear instructions for what each manual segment is to include, and the final coordination and editing of the completed manual.

Team leaders should have strong writing and editorial skills because they will have the job of making language, style, and format internally consistent. A good team leader will make use of instructions, writer guidelines, writer checklists—any procedure that helps writers know what is expected of them, and when. In the last section of this chapter, we provide some samples of checklists, work schedules, and information-gathering techniques. These may be adapted to match your company's structure and procedures.

What the Writer Needs

Writers, whether in small companies or large, have some very basic needs. To do a good job, they must have access to information and adequate time.

Information

Believe it or not, we have met many writers whose chief frustration was a lack of information about the product. They may ask to see the product and be refused. They may ask for scheduled time to review the product with designers, technicians, engineers, or safety personnel and be told that there is no time. They may ask for a working model, a prototype, or at least a photo and get a flat no for an answer. Or they may be housed in an office miles away from where the product is produced. Admittedly, many people may be clamoring for a prototype of a new product. Marketing wants it, engineering is working on it, product safety needs it. When the pressure is on and deadlines must be met, writers often get short shrift.

However, if the manual is to perform its function, writers must have information, and management must provide the procedures to help them obtain it. Information gathering is an important first step in planning the manual.

Time

Our comments on time as a basic writer's tool are directed especially to managers. Deadlines are the name of the game in most industries. More errors and slapdash jobs can be explained by time pressures than by incompetence. Managers often need to be reminded that writing takes time. Writers themselves usually do not have to be convinced. They know that writing takes more time than anyone would ever guess, though they sometimes underestimate how much. (This book, for example, was a project undertaken in addition to our teaching jobs. The information took several years to collect, and the book took a year to write, plus months of revision and rearrangement.) We have surveyed writers in our seminars who estimate that even an average one- or two-page business letter or memo may take several hours to compose.

Once writers and/or supervisors have created effective information collection systems and have gone through the manual production process at least once, subsequent manual production proceeds more quickly. But totally new products need especially generous lead time for creating the manual, since some of the vital information may not be available until the last minute, when the prototype is completed and tested.

Information: How to Collect It

After you have decided on the best allocation of personnel, time, and money, you can begin the initial planning of the manual itself. At the planning stage, it is essential to gather information and establish schedules, deadlines, and writer guidelines.

Collecting information about a product may be the responsibility of one writer, the writing team leader, or the publications supervisor. Quite often, the responsibility is shared. Companies vary a great deal in the procedures they use for information gathering. In a very small company, the writer may need only to lean across the desk and ask a co-worker for information. In very large companies, information collection is usually more complicated and is often accomplished by means of scheduled meetings, writer checklists, and strictly regulated systems of sign-off and review.

Adapt your information-gathering system to the realities of your company, but by all means do not attempt to begin writing before you have essential information in hand.

Information collection consists, in part, of making sure the right people are able to talk to one another and, in part, of having essential technical data in hand. Listed below are some of the successful techniques used by industries:

Information-Gathering Aids

- Regularly scheduled conference time between writers and engineers, designers, and/or technicians.
  (This time need not be long if both parties are prepared with basic information and a list of key questions.) Engineers and designers should expect to spend this conference time with writers on every product.
- A product safety committee that includes writer representatives when key safety features and messages are being discussed.
- Placement of writers' offices near production, research, and test facilities.
  This assures that the writers are not working in a vacuum. (One look at a product is worth 20 phone calls.)
- An orderly file system that allows the writer to reuse materials prepared for other manuals, especially if the manual describes a new product with only minor design changes or a slight model change.
- A product history that alerts writers to places where bad manual writing may have caused operator problems, accidents, or death.
- A scheduled walk around the product or prototype well in advance of the manual deadline.
- Writer guidelines and style handbooks prepared in-house by editors.
- A writer's checklist before manual work begins that includes these items:

  —specifications and dimensions

  —brief description of product function and use

  —important safety features and hazards

  —new design features unfamiliar to the writer

  (This information is provided by engineering, marketing, product safety, testing—whatever group bears responsibility or has the information.)

All the suggestions listed above are subsumed under a single prerequisite: *writers must have access to information.*

Time: How to Schedule It

Manual production becomes much easier if, at the planning stage, you can determine approximately how much time you can devote to each step of the production process. A product is rarely allowed to leave the production facility until the accompanying manuals or instructions are complete and ready for distribution or packaging with the product. The controlling date, then, is the deadline when the product is scheduled to be shipped or sold. Establish this fiinal deadline, and create a work-flow schedule that allots time for the following steps:

Phase 1:   Initial Planning
- Writer assignments
- Writer checklists and guidelines
- Information collecting
- User analysis
- Outline development

Phase 2:   Preparing the Manual
- Writing and layout
- Creation of visuals
- Reviewing
- Editing
- Revision
- Preparation of final copy
- Printing

Steps of Manual Production

The following sections describe the activities involved in each of the steps of manual production.

Phase 1:   Initial Planning

Writer Assignments

These may be for a complete manual or for segments of a manual. Decide whether to use solo writers or writing teams, and assign responsibilities for editing the manual and for making final judgments on binding, paper stock, and page size.

Writer Checklists and Guidelines

These may be prepared by the writer(s) and/or supervisors. Guidelines include lists of specifications and dimensions, special safety hazards, and new design features of the product. These may also include standard in-house instructions for format, standardized glossaries of terms for parts and procedures, and the types and sizes of photos and drawings that may be used.

Information Collecting

Sources of information may be product designers, engineers, and personnel in marketing, sales, research and development, product safety, and production. Other sources may include product histories, files, photos, test data, and reusable materials from other manuals.

User Analysis

Steps for user analysis are described in Chapter 2, "Analyzing the Manual User." Analyze the user before writing the manual—this will allow you to decide on appropriate format and language levels for the manual. User questions may be employed to develop the outline and to determine what major sections the manual should contain.

Outline Development

Develop outlines for major sections and/or chapters of the manual. Outlines should include all relevant technical information acquired during information collection. Employ user questions (see Chapter 2) to establish major sections. Use outlines to refine cross-referencing between manual sections and to prevent redundancy and overlap.

Phase 2: Preparing the Manual

Writing and Layout

With essential product information, writer guidelines, a clear idea of potential manual users, and fully developed outlines, writers can begin writing and layout.

Creation of Visuals

Creation of visuals goes hand in hand with writing and layout. Frequently a photo, drawing, or chart will be created first, and the accompanying verbal text will then be written to back up the visual. For some manuals, the visuals *always* come first, and text is developed from them.

Reviewing

Small companies often conduct continuous informal review as the manual is being produced. Large companies or companies with complex products usually have more stringent and codified procedures for sign-off and review. Reviews may include

- Technical review (by designers, engineers, etc.)
- Legal review (by legal staff or counsel)
- Product safety review (by safety personnel and human factors engineers)

There may be single reviews and sign-offs, or multiple and repeated reviews. Technical, legal, and safety reviews may come early or late in the manual process, depending on the company.

Editing

Editing may be done once or many times and must be coordinated with the other reviewing procedures.

Revision

Revision goes hand in hand with editing. It may be done once or many times, depending on company structures (i.e., solo writers, teams, supervisors) and on time constraints. Revision should include debugging with a person-on-the-street user. (See Chapter 2.)

Preparation of Final Copy

Final copy should be camera-ready for printing.

Printing

Allow adequate lead time if printing is done by outside contract.

Phases of Manual Production

Keep in mind, as you plan the steps of manual production, that the work flow described in our schedule is a two-phase process.

Phase 1 is often squeezed out or omitted entirely when deadlines are pressing. Remember that time spent on initial planning, user analysis, and outlining is not time wasted, but time gained. Phase 1 sets up the work place and gives writers the essential tools they need for Phase 2. If you lay the groundwork in Phase 1, you will gain back the time spent when you reach Phase 2.

Phase 2 is seldom a series of smooth steps. Steps will overlap and will vary in their sequencing, depending on the size of the company and its organizational structure. For example, some companies have not one but several review or editing steps that may come early or late in the work-flow schedule. Further, the creation of visuals, the planning of format, and the writing of verbal text usually occur simultaneously. Finally, the best of schedules will develop glitches—missing data, late photos, key people sick or out of town, last-minute design changes.

In short, you need to sequence and overlap the steps we suggest in a way that makes sense for your company, but do not neglect or omit Phase 1. (See Chapters 3, 4, and 5 on writing strategies, formatting, and visuals.)

Scheduling Responsibility

In large companies, manual scheduling is usually handled by team leaders or publications editors. If you are working as a solo writer, you will have to do much of the scheduling yourself. Schedule yourself extra time when you have to collect information from other people or when you have arranged for outside help (e.g., with typing, printing, or drawing).

Summary

Planning is the first step to successful manual production. It begins with intelligent allocation of time, money, and personnel and with organization of the work place. Writers need information and time. Do all you can to fulfill these basic needs, and regard information collection, user analysis (Chapter 2), scheduling, and outline development as the essential groundwork for manual writing.

2

Analyzing the
Manual User

Overview

User analysis should be included in initial planning of the manual. (See Phase 1, "Initial Planning," in Chapter 1.) Writers should have a clear picture of manual users and their needs. What the user needs to know should guide manual writers in choices of language level, reading level, safety warnings, execution of visuals and graphics, and arrangement of manual segments—all the elements discussed in subsequent chapters of this book.

This chapter gives you guidelines for user analysis and user feedback. We describe the spectrum of manual users and show the differences between the professional and the general public user—although this chapter focuses on the general public user. We do this for two reasons:

- Manuals for general public users are the most common.
- Many of the guidelines for general public manuals also apply to manuals directed at professionals. (Chapter 7 treats the special problems of service manuals, which are usually written for professionals.)

We provide a list of products most likely to be bought by the general public and some questions to use as guides to analyzing your user. The last segments of the chapter suggest techniques for collecting user feedback. We show you how to simulate the person-on-the-street user as a way of avoiding "shop blindness''—the inability to see your products as a first-time user might—and we show you how to make use of feedback in revising and updating manuals.

At the end of the chapter, you will find a checklist to help you analyze your user.

Know Your User

To work effectively, a piece of writing, just like a piece of equipment, must be designed with two questions in mind:

1. Who is supposed to use it?
2. What is it supposed to do?

Until writers answer these questions, they cannot arrange their material effectively.

The answers to the first question, Who is supposed to use it? determine many of the answers to the second question, What is it supposed to do? For example, if the user of a manual is a first-time buyer of a riding mower, the writer of the manual must not assume that the user will know how to maneuver the mower on a hill. On the other hand, if the user of an electrical service manual is an experienced electrician, the manual writer can assume that the user will have a basic knowledge of electrical circuits.

Spectrum of Users

The wide range of potential users poses a problem for manual writers. At one end of the spectrum are the professionals; at the other, the general public. As a result, when you first ask yourself, "Who is supposed to use this manual?" you may only envision an undefined, anonymous buyer. Let's take a look at some definitions of *professionals* and *general public* and at some user characteristics and distinctions. These should help you to bring into sharper focus the hazy picture of your manual user.

Professionals

Certain kinds of manuals and instructions are written expressly for professionals—for example, the service manual for the trained mechanic or the instructions for a surgical probe.

Definition of a Professional   A professional may be defined as a user who is likely to have special or in-depth knowledge of the product, its purpose, and the technical terms used to describe it.

For example, some manuals are written exclusively for such professionals as these:

- Licensed electricians or plumbers
- Registered engineers (civil, mechanical, nuclear, chemical, metallurgical, industrial)
- Trained and licensed service and repair persons
- Master carpenters or mechanics
- Medical professionals (doctors, nurses, lab technicians)
- Computer operators (designers of software or hardware, computer maintenance personnel)

Manuals for the professional are often easier to write because you can make certain safe assumptions about the user's understanding of the product. That confidence is usually reflected in the level of language you are free to use in talking to the professional.

Example 2.1 contains two passages that describe the same mechanism. The writer has manipulated several language elements, adjusting the passage with specific users in mind.

Comment   Both passages describe the same mechanism. Passage A assumes the reader knows the basics of pump operation and venting and also knows terms like *output variation, load variation*, and *torque* without having them defined. Passage B uses a minimum of technical language and explains the venting action by means of the simple analogy of the soda straw.

Quite often, the same product will have two manuals, written at different language levels. Or it may have separate manual segments, some written for professionals, some for general public users. These two passages could both occur in the same manual—the first in a segment directed to surgeons or trained technical personnel, the second to equipment assistants and nontechnical personnel.

(The service manual for professionals relies on the user's in-depth knowledge of the product but also demands that the manual be put together somewhat differently than a manual for the general public. We deal with this special kind of manual in Chapter 7, "Service Manuals.")

Guideline   Manuals for professionals may safely use more technical language and visuals but must be as clear and logical as manuals for the general public.

Example 2.1   Manipulation of Style with the User in Mind

Passage A—for the Professional

A peristaltic pump is used to create the suction for vacuum. You can use different vacuum levels for the various handpieces and "tips" employed during surgery.

A constant volume peristaltic pump delivers a constant flow rate of 28 + 2 cc/min. The pump is driven by a regulated DC control voltage that has less than 2% output variation over AC line variation of 106 to 128 VRMS and a load variation of 0 to 75 ounce-inches of torque.

Passage B—for the General Public User

Venting action causes the vacuum level at the port to drop to less than 2 inches of water in less than 300 milliseconds and occurs automatically each time the foot switch moves from Position 2 to Position 3.

An important part of the vacuum system is the vent control. A simple way to understand venting is to think about what happens to fluid in a soda straw. The fluid is sucked up into the straw. (This is analogous to the vacuum buildup taking place in the ABC system.) If you place your finger over the top of the straw, the vacuum is maintained, and the fluid stays in the straw with little or no leakage. To release the fluid, remove your finger from the top of the straw and allow air to enter. In essence, this is what happens in the vacuum system of the ABC machine.[1]

General Public

Far more common are manuals for the general public (sometimes called consumer manuals). The term *general public* seems at first glance to be so broad that it defies definition. Begin by looking at the list below, which gives an idea of the categories of products usually intended for the general public user.

Typical General Public (Consumer) Products

1. Appliances
2. Automotive products
3. Biomedical devices (prostheses, blood pressure kits, contraceptives, heating pads, trusses, braces, contact lenses and glasses, orthopedic aids, hearing aids, dentures)
4. Construction equipment
5. Drugs and health products
6. Farm and industrial equipment
7. Firearms
8. Foods
9. Household products (soaps, polishes, cleaning agents, sprays, pesticides, stools, ladders)
10. Office equipment
11. Paints, general-purpose chemicals (fertilizers, solvents, removers)
12. Power and hand tools
13. Sporting goods (bicycles, minibikes, skis, swimming pools)
14. Toys

Despite the breadth of the term *general public*, the general public user may nevertheless be defined according to the following characteristics.

1. *Biological Characteristics*. Male or female, of any age

A general public user can be a male or female of any age, from child to oldster. If you write instructions for skis designed for the resilient 16-year-old body, you have no assurance that a 40-year-old won't try them. If you make a caustic toilet bowl cleaner for a housewife to use, you have no assurance that a preschooler (who can't read the label) won't think it's just another colored powder. Increasingly, many products formerly intended solely for one sex or the other are being used by both—such products as blenders, power saws, shotguns, and hair dryers.

2. *Literacy*. Illiterate to college-educated

Illiteracy and declining literacy are modern realities. Instructions for use by the general public should not assume literacy. If the product is especially complex, unusual, or hazardous, reliance on visuals or pictures is essential, in the manual or in instructions affixed to the product. (The special problem of safety warnings is discussed in Chapter 6.)

3. *Technical Sophistication*. Sophisticated to naive

The general public user may be technically sophisticated or technically naive. If your product is intended for general use, you must aim the manual at the naive user. The technically sophisticated will be able to follow along anyway and will simply be able to use the manual more quickly. If the manual aims for the technically sophisticated user (and many manuals make this error), the naive or inexperienced user is left behind. The manual is then useless to that reader and goes unread.

User Characteristics and Distinctions

Most manufacturers, by knowing the market for their products, can aim or target manuals, written instructions, and warnings by using the following list of user characteristics and distinctions. The list gives you more information for deciding if your users are professional or general public.

1. Personal characteristics of the user
   - Does he or she use this machine or product almost every day or only once in a while?
   - Is he or she likely to have or have used other products like it?

- Does he or she do his or her own routine maintenance?
- Does he or she do his or her own repairs? Should he or she?
- Does he or she understand technical language?
- Does he or she understand charts, circuit diagrams, mechanical drawings?

2. Conditions of manual use

- Will he or she use the manual only to learn how to set up, use, or operate the product?
- Will he or she refer to the manual or instructions often?
- Will he or she use the manual or instructions only if something breaks, fails, or appears to be abnormal?
- Will he or she read the entire manual or only a section here and there?
- Will he or she be able to look at the machine or product when he or she is reading the manual?
- Will the light be good?

3. Information wanted

- Basic instructions for use, operation, and adjustment?
- Routine maintenance procedures?
- Sophisticated service procedures?
- Specifications and parts lists?
- Troubleshooting procedures?

The same user may respond differently to different products. If, as a writer, you believe that you are writing too simply or too nontechnically, consider how the following situations demand simplicity and nontechnical clarity in written instructions:

- The housewife trying a new cooking oil? Repairing a faucet? Replacing spark plugs? Using a power saw?
- A 16-year-old boy repairing his bike? Making babyfood?
- An 80-year-old man learning to use a computer? A power mower? A microwave oven?

None of these situations is impossible or even improbable. A novice in one situation may be a professional in another.

Guideline   General public manuals should be simple, clear, and nontechnical.

User Questions as Organizers of the Manual

After you have established a clearer idea of who is supposed to use your manual, you are ready to determine what the manual should contain and how its major sections should be organized. Ask yourself, "What is the manual supposed to do?" Your user can be your guide here. Lay out the manual sections as if they were responses to user questions. To do this, you must step back from your product and try to see it as first-time buyers or users might. What questions will they ask?

- "Interesting-looking toy—how does it work?"
- "Sunglo—I heard about this on TV. Let's see what's in it to make the windows shine. Do I have to mix it with water? If so, I'm not interested."
- "Crazy-looking copying machine. Where do you put the paper? How do you stop it?"
- "I could get this done a lot faster if I used my dad's power saw. How do you feed the board into it?"
- "Is it dangerous to use my electric razor while I'm sitting in the bathtub?"

Imagine the buyers of your product shopping in a store or dealer showroom. They pick up the product, talk to the salesperson, or read the manual of instructions for its use. They do this because they have questions. Answers to those questions can form the major sections of your manual.

Look at this chart to see how a user question can form a section of your manual. The column on the left gives you a typical user question and its corresponding manual section. The columns to the right contain typical answers for two products, an exterior paint and a microwave oven. These answers could be used to make up the information contained in major sections of the manual.

Manual Section Derived from User Question	**Answer to User Questions about the Product**	
	Exterior Paint	*Microwave Oven*
Scope (What is the main function of this product?)	One of 25 products for protective exterior coating.	In combination with conventional stove.
Description of Product (Is this what I'm looking for? Introduce me to it.)	Oil base, high quality, for residential or commercial buildings. Fifteen colors available.	Home unit for cooking/heating food.
Theory of Operation or Intended Use (How does it work? What is it for?)	Protective coating for wood, asbestos, brick, stucco.	Home food preparation.
Special Features or Design Details (What is special about it?)	Durability, nonfade color, controlled replacement color formula.	Reduces preparation time.
Limits of Operation or Use (What are its limitations?)	Not for metal, glass, plastic. Brush application only—not spray. Mix only with organic solvents; no water.	Use for cooking only. Door must be closed.
Setting up/Turning on (How can I assemble it? Turn it on?)	Brush application. Temperatures above 50°.	Professional installation. Explanation of controls.

Manual Section Derived from User Question	Answer to User Questions about the Product	
	Exterior Paint	*Microwave Oven*
Normal Operation or Use (What is normal use and life of product?)	Dries in 24 hours. Seven-year life. Can be washed.	Timetable and instructions for food.
Turning off/Disposal (How do I turn it off? Dispose of it?)	Excess and accompanying solvents flammable. Precautions for handling.	Turn-off controls.
Abnormal Operation (What tells me something has malfunctioned?)	Causes of cracking, peeling. Not for internal use.	Signs of malfunction. Safety features.
Preventive Maintenance (How do I take care of it?)	Proper surface preparation. Proper application.	Cleaning. Professional check.
Storage (How do I store it?)	Close lid tightly. Store upside-down. Shelf life.	
Safety (How do I use it safely?)	Safety information will be found throughout the manual and on the product. See Chapter 6, "Safety."	

Users as Feedback Sources

As we have shown, your manual user is important as the audience and organizer of your manual. The manual user can also be helpful in providing feedback, especially at the revision and follow-up stages of manual production. Writers who rely on user feedback tell us that this feedback step is invaluable for debugging a manual before the final copy is printed and for assessing manual use and effectiveness after the product is sold.

Simulating the User: The Person on the Street

Overfamiliarity with your product can lead you to make the assumption that the user knows as much about your product as you do. One effective way to find out if you have avoided the "shop blindness" that comes from knowing your product too well is to submit your manual to a person-on-the-street test. Invite an employee from another division, a secretary, your child—anyone who is a true stranger to the product—to "walk through" the manual, following its instructions and descriptions.

Stand by and listen, but don't provide verbal backup to the manual unless the user asks for help. Wherever you have to break in, explain more fully, or provide more information, your manual probably needs revision or clarification.

Some companies use a hidden camera to check manual effectiveness. They recruit a willing person-on-the-street user from inside the company or from the general public, put the manual user alone in a room with the product, and film the user as he or she assembles and operates the product. The film helps manufacturers to see exactly where users go wrong in following instructions.

User Interviews and Surveys

Many companies now do follow-up surveys and field checks on their manuals by asking users for feedback. Some companies simply include a postage-paid card in their manuals asking users to assess manual effectiveness. (Response in this way is likely to be low unless the manual is very bad.) Others send out company troubleshooters who contact buyers, dealers, and service technicians and ask for feedback on whether manuals are working well or need revision.

Manipulating Manual Elements to Match User Needs

The information you gather about your users should be your guide as you make choices in writing the manual. User analysis affects:

- Language level and reading level
- Choice and execution of visuals and graphics
- Proportion of visual to verbal text
- Arrangement of segments of the manual
- Safety warnings
- Revision and updates

These elements may be manipulated in a number of ways to match user needs and questions. (See subsequent chapters for fuller treatment of visuals, graphics, and safety warnings.) Suppose, for example, that you are considering how to manipulate language levels and reading levels to accommodate users you have identified as either general public or professional. Example 2.2 shows how manual writers have controlled reading and language level.

Example 2.2   Two Passages Showing How User Analysis Affects Language and Reading Level

Below are two different descriptions of the same procedure (removing front brake shoes from standard drum-type automotive brakes). The first is from a Dodge service manual, the second is from *How to Keep Your Volkswagen Alive: A Manual of Step by Step Procedures for the Compleat Idiot* by John Muir.

From the Dodge Manual

1. Using Tool C-3785 remove secondary return spring then remove adjusting cable eye from anchor. (Note how secondary spring overlaps primary spring.)

2. Remove primary return spring.

3. Remove brake shoe retainer springs by inserting a small punch into center of spring and, while holding backing

plate retainer clip, press in, and disconnect spring. Un-
hook cable from lever. Remove cable and cable guide.

4. Disconnect lever spring from lever and disengage from
   shoe web. Remove spring and lever.

5. Remove primary and secondary brake shoe assemblies
   and adjusting star wheel from support. Install wheel
   cylinder clamps (Tool C-416) to hold pistons in
   cylinders.[2]

From the Muir Book

Look at the brake assembly; you'll find there are return springs
holding both ends of the shoes toward each other. Look at how
they're fastened—see how they're anchored. You'll have to
replace them the same way. Clamp the vice [sic] grips on the
closest spring to you then pry it out of its hole with the
screwdriver. Remove the other springs the same. Take out the
two round springs with caps over them in the center of the
brake shoe webs, holding the shoes to the brake plate. They'll
come off with your fingers, so hold the pin in the back of one
with your forefinger and push and twist on the little cap with
the thumb and forefinger of the other hand. When that little cap
is 90° around on the pin, it will come off and the whole thing
will come apart. Take a close look, you have to put them back
on. . . . Remove the other spring-cap-pin assembly from the
other side. Now you can work them out of their slots. Snap a
rubber band tight around the wheel cylinder slots so the wheel
cylinders won't come apart.[3]

Comment   The procedures described in these two passages
are identical, but the writers had two very different kinds of
users in mind. Strengths and weaknesses of the two approaches
include the following:

Dodge Manual

- Written for a professional mechanic.

- Assumes familiarity with tools, parts, and mechanisms.

- Numbered list format makes it easy to follow.

Muir Book

- Written for novice mechanic, do-it-yourselfer.
- Anticipates trouble spots (especially those requiring unusual coordination of hand or hand and eye).
- Warns of disassembly problems.
- Tries to use common terms for tools and parts.
- Harder to follow than Dodge manual because of paragraph format instead of list.
- Chattiness makes it wordy, but still more appropriate for novice.

Feedback from users can also help you to make adjustments in manuals by revising and updating or by altering manuals as successive models of a product come out. Example 2.3 shows how user feedback from a survey can sharpen insights on actual use of the manual.

Example 2.3   Feedback from Manual Users: Surveys

> Talking with manual users, either informally or through structured surveys, can be informative and can give you insights on the strengths and weaknesses of your manuals.
>
> To see what kind of feedback might be provided by manual users, we conducted a series of informal interviews with service people, dealers, and buyers of agricultural equipment. To encourage honest responses, we tried to ask questions that would encourage people to talk about areas of the manuals that most concerned them. We asked, "Do you use the manuals for your equipment? If so, how? If not, why not?"
>
> We found that dealers and service people answered the questions somewhat differently than buyers (farmers). Here are some of the responses to our questions. Notice, as you read, that dealers and service people stressed the importance of the manual as a teaching tool and a legal document and that they were keenly aware of different levels of technical sophistication among buyers. Farmers had occasional positive things to say but had many complaints about manual effectiveness.
>
> At the end of the user responses, you will find our comment on what we learned from this feedback exercise.

Informal Survey on Manuals for Agricultural Equipment

Dealers and Service People

1. *Levels of Sophistication*

   - "There's a big difference in some of these modern-day farmers. There'll always be some who think the manuals are too complex—they want to be told to tighten down a nut, not what its dimensions are. On the other hand, if they're used to checking on a corn planter just by eyeballing it and then they buy one with electronic equipment, they'll sit down and read every word."

   - "No use lecturing a farmer about stuff he's used all his life. He tends to think all engines are pretty much the same. He knows how to drive a tractor; he's been doing it all his life. He'll look at the manual if there's a new wrinkle, but he has to have it pointed out to him."

   - "I guess there's a lot of the 'good old American know-how' in most of my buyers. I get my calls when they've taken something apart or tried to fix it and can't. If it looks like they've botched a job and the labor bill is going to get bigger if they go on with it, then they'll call us to come out or bring it in."

   - "Some of my buyers are buying for as many as 32 farms. They've got full service departments to take care of their equipment—some of their mechanics are better than ours. They're very well educated, and they don't want to be talked down to."

   - "The ones that need the most elementary help are these 90-day wonder suburbanite farmers with 15 acres of land. They're probably buying a small tractor for the first time—everything's new to them."

   - "You could mount the safety instructions in neon lights on the power takeoff, and 50 percent of the old-hand farmers will remove whatever gets in their way."

2. *The Importance of the Manual*

   - "It used to be that the operator manual got filed away or tossed somewhere. Now they're getting more and more use. The equipment is more complicated, more things can go wrong, more malfunctions are adjustment problems.''

- ''Our service and delivery people use the manuals for teaching. In fact, we set up a school for all our buyers of corn planters, bailers, and combines. The service people run those schools, and they show the buyers what kinds of adjustments and troubleshooting are likely to present problems. They use the manuals to show a farmer what he can do for himself before he calls for help.''

- ''We think the manual is so important that we require dealers to register the serial number of the manual as well as the equipment.

  1) We want to make sure he's got it and have proof of it.

  2) We want to emphasize its importance to our customer.

  3) Liability settlements are getting bigger all the time—there are more and more ways to get hurt.''

- ''One of the biggest complaints is that the manuals don't go deep enough. The pictures are fuzzy, or they don't know what kind of tool they need.''

- ''You've got to keep the servicing for agricultural equipment separate from consumer lines. A guy with a $50,000 piece of equipment does *not* want to stand around all Saturday afternoon waiting for an answer while your service people are working on something that cost $500. Especially in the spring. Most farmers will cobble things together any way they can to keep going during planting or harvesting. In January or February, they might be more leisurely about things.''

Farmers

3. *Mixed Responses*

- ''The pictures are terrible. Half of them are too small and full of those damn little numbers. You can't see what the arrows are pointing at, and you have to keep flipping back and forth. Most farmers I know have eyesight that's none too good—all that crap flying around in the air, you know.''

- ''You ought to talk to about 500 farmers. I'll bet they'd all tell you those things are put together by a bunch of engineers sitting in offices with fluorescent lights. Most farmers aren't engineers. They're working in mud or in a dark tool barn.''

- ''I weld a lot. I know there's some stuff I shouldn't, 'cause when I try to turn or something, I find out what I welded doesn't work so good anymore—wasn't made to be welded. But some of my stuff is pretty old, and a trip to the dealer is 35 miles and a five-hour ordeal, and then I might not even get the part. Who's got that kind of time?''

- ''I wish they'd always give you the manual that goes with your equipment. Some of them don't really match the machine you've got. I can't tell you how many times I've stared at those pictures and wondered if that funny little thing sticking out on the picture is the same as the funny little thing on my machine.''

Comment   Given a chance to talk about how they used manuals, these farmers, dealers, and service people had good and bad things to say about them. The most common responses tended to fall into the categories you see, although questions and conversations themselves were informal.

From these responses, we learned that

- Users are general public (range from technically sophisticated to technically naive).
- Manuals for products with new design features or repeated service problems get more use.
- Confusing visuals are a problem for many users.
- Users sometimes are given the wrong manual or a generic manual intended to serve for several models of a product.
- Dealers and service people rely heavily on manuals and sometimes use them for teaching.

- Conditions for manual use are sometimes bad (field, mud, barn).
- Farmers tend to use manuals for doing breakdown, maintenance, repair, and setup or for understanding new design features.
- Farmers work under pressure and do much of their own service and repair.
- Farmers probably seldom read the whole manual for a product they are familiar with.

Checklist: User Characteristics and Distinctions

Answer these user questions about the manuals you and your company produce. After you have answered the questions, consider whether you would want to change anything about the way your manual is put together.

1. Personal characteristics of the user
   - ☐ Does he or she use this machine or product almost every day or only once in a while?
   - ☐ Is he or she likely to have or have used other products like it?
   - ☐ Does he or she do his or her own routine maintenance?
   - ☐ Does he or she do his or her own repairs? Should he or she?
   - ☐ Does he or she understand technical language?
   - ☐ Does he or she understand charts, circuit diagrams, mechanical drawings?

2. Conditions of manual use
   - ☐ Will he or she use the manual only to learn how to set up, use, or operate the product?
   - ☐ Will he or she refer to the manual or instructions often?
   - ☐ Will he or she use the manual or instructions only if something breaks, fails, or appears to be abnormal?
   - ☐ Will he or she read the entire manual or only a section here and there?

☐ Will he or she be able to look at the machine or product when he or she is reading the manual?
☐ Will the light be good?

3. Information wanted
☐ Basic instructions for use, operation, and adjustment?
☐ Routine maintenance procedures?
☐ Sophisticated service procedures?
☐ Specifications and parts lists?
☐ Troubleshooting procedures?

Checklist: User Questions as Organizers of the Manual

This checklist can be used either to plan and outline a new manual or to evaluate a manual you have already written.[4]

	User Questions	**Segment of Your Manual That Answers Question**
1. Scope	1.	1.
2. Description	2.	2.
3. Theory of Operation	3.	3.
4. Design Detail	4.	4.
5. Limits of Operation	5.	5.
6. Setting up and Turning on	6.	6.
7. Normal Operation	7.	7.
8. Turning off	8.	8.
9. Abnormal Operation	9.	9.
10. Preventive Maintenance	10.	10.
11. Storage	11.	11.

Summary

Be aware of the manual user as you plan your manual. The following steps will sharpen your awareness of the manual user and help to organize the manual.

- Think systematically about your manual user.
- Anticipate user questions.
- Employ answers to user questions to structure the manual.
- Employ user feedback to revise, refine, and update manuals.

References

1. Adapted from a manual by CooperVision Systems, an operating unit of CooperVision Surgical, a division of CooperVision, Inc. Used by permission.

2. *Dodge Dart, Coronet and Charger Service Manual 1967.* Detroit, MI: Dodge Division, Chrysler Motors Corporation, 1967, pp. 5–6. Used by permission.

3. Muir, John, and Gregg, Tosh. *How to Keep Your Volkswagen Alive: A Manual of Step by Step Procedures for the Compleat Idiot*. Santa Fe, NM: John Muir Publications, 1974, p. 157. Used by permission.

4. Checklist adapted from material developed by Emeritus Professor Charles A. Ranous, University of Wisconsin. Used by permission.

3

Organization and Writing Strategies

Overview

People tend to acquaint themselves with an unfamiliar product in much the same manner as they find their way around a new city. They look at maps and search for landmarks and street signs. As you develop a manual, you can simulate the function of street signs and landmarks by careful organization and by writing strategies that help readers find their way around the manual and feel comfortable with the product.

This chapter shows how to organize and write manual sections, which may be several pages, a cluster of several paragraphs, a single paragraph, or a single sentence in length. The following are the most commonly used organizing strategies.

1. General headings, followed by one or a number of pages that fall under that heading
2. Paragraph clusters, usually appearing as subheads under a general heading
3. Single paragraphs organized by the presence of a heading, a core sentence, and/or supporting sentences
4. Single sentences organized by various strategies, such as lists, parallels, series, comparisons, cause-and-effect statements

Organization of the manual also requires that you pay attention to the sequences of segments. We discuss approaches to sequencing: general to specific chronology, and spatial logic. The last part of the chapter shows you strategies to reduce the length of verbal text. Throughout the chapter, you will see examples of manual pages and, at the end of the chapter, some sample pages and questions that can be used for review.

General Headings: Chapters and/or Major Sections

We use the term *general headings* to denote those headings used for identification of manual chapters and/or major sections. General headings provide landmarks for user understanding of major systems of a machine, major uses of a product, and/or major steps in a process. As suggested in Chapter 2, you can employ user questions in determining the major sections of a manual. We stated that those major sections would include:

Scope

Description

Operation and Intended Use

Special Features

Limits of Operation and Use

Setting Up and Turning On

Normal Operation

Turning Off and Disposal

Abnormal Operation

Preventive Maintenance

Storage

Safety (throughout the manual)

Most manuals will have some, if not all, of these sections, and you may choose to vary the number, order, and length of major sections to match the requirements of your product. The following are some examples of the general headings used to designate chapters or major sections of manuals for a variety of products.

Examples of General Headings

Surgical Aspirator
Scope and Purpose of Manual
Introduction to Model 923
Description and Function of
   Handpieces
Description and Function of
   Console
Unipaks
Instrument Setup

Herbicide
Precautions
Uses of Product
Mixing and Spraying
Fluid Test Compatibility
Application Information
Cultivation Information
Weed Control

Camping Tent
Capacity of Tent and Uses
Parts of Tent
Assembly
Disassembly
Care of Tent
Storage

Fabric Dye (home use)
Fabrics Suitable for Dye
Preparation of Fabric
Mixing and Preparing Dye
Applying the Dye
Fabric Care and Washing

Shower Fitting
Tools Needed
Specifications
Replacement Installation
New Installation
To Remove Valve Assembly
To Replace Valve Assembly

Tractor
Safety
Controls and Instruments
Pre-start Checks
Operating the Tractor
Drawbar and PTO
Ballast
Transporting
Wheels, Tires, Treads
Lights and Signals
Fuels and Lubricants
Lubrication and Maintenance
Service
Storage
Troubleshooting
Lubrication Chart

The examples show both chronology and spatial logic and, of course, reflect the complexity of the product. For instance, the herbicide manual is a small, pocket-size booklet of 45 pages, the tractor manual is 102 pages, and the tent manual is 8 pages. As the length of the manual increases, the general headings become more important as road signs for the user. And for you, the writer, the general headings are invaluable as you develop the outline for your manual.

Headings as Hierarchical Cues

Headings not only tell the reader the main idea of the section to follow, but also provide cues on the relative importance of various kinds of information. These cues are largely picked up by responses to print size, type face, uppercase or lowercase letters, and position of the heading on the page. Individuals may disagree here and there on the meanings of these cues, but, in general, these responses hold true:

1. Largest level of organization or most important information
   - larger type
   - bolder print
   - all uppercase letters
   - centered on the page

2. Smaller level of organization
   - smaller type
   - finer print
   - mix of uppercase and lowercase letters, or all lowercase
   - flush left or indented

3. Special emphasis
   - italics or underlining
   - boxes
   - color

For example, a section of a manual laid out in the following way will be perceived by the reader to be arranged in a descending hierarchy of ideas:

Controls

On-Off

Pre-set

Throttle

Turning

- right
- left
- circle

If you are having your manual produced by professional printers or by in-house typesetting, you have many levels of hierarchy available to choose from (typefaces, print size, upper-case and lowercase). However, even if you are producing your manuals simply by typing them in the office, you still have the choice of all uppercase, all lowercase, or a mixture of upper and lower, plus underlining. If you set up all your headings in one mode only (such as uppercase), you are wasting valuable design options that are available merely by varying the choice of modes.

Look at Figure 3.1, "Effective Use of General Headings," which shows the power of print size, uppercase and lowercase letters, boxes, and color.

When manual users see a page like Figure 3.1, they perceive the following hierarchies:

1. General Heading: STORAGE

   - The fifth major section of the manual.
   - Separated from the bottom text by full-page under-score line.
   - "Storage" is repeated at the bottom right of the page in italics and is repeated, in the same position on the page, on all pages that belong to this major section.
   - Very large "outline" type indicates further that "Stor-age" is a major section.

Figure 3.1.   Effective Use of General Headings. This figure shows the power of print size, uppercase and lowercase letters, boxes, and color.

---

5 Storage

STORING THE TRACTOR

Preparation

1. Change hydraulic oil.

2. Change transmission oil.

3. Change engine coolant.

4. Drain and flush gear oil from differentials and planetary gear housings. Fill with new oil.

5. Change engine oil and filters.

6. Start engine. While engine is warming up, operate the transmission, hydraulic system, steering and differentials to distribute the new lubricants to components. Warm engine to at least 70° C (160° F); it may be necessary to shield the radiator to achieve this temperature. Stop engine.

7. Clean tractor of all debris, dirt and accumulated grease.

8. Drive tractor to storage location.

9. Relieve tension on alternator, air conditioner compressor and fan belts.

10. Coat all exposed hydraulic cylinder shaft areas with grease or a rust preventive.

Storing

1. Using plastic bags or tape seal the following openings: air cleaner inlet, exhaust muffler, fuel tank breather and air conditioner air intake screens.

2. Touch up all scratched or chipped painted areas.

3. Block up tractor to remove weight from tires.

4. Cover tires if they will be exposed to heat and/or direct sunlight.

5. If tractor is to be stored outside, cover with a waterproof canvas or other protective material.

STORING BATTERIES

⎡ WARNING ⎤

⚠ AVOID SMOKING OR OPEN FLAMES IN OR NEAR BATTERY CHARGING AREA DURING OR FOR TWO HOURS FOLLOWING CHARGING.

BE ALERT

1. Low maintenance batteries do not require charging before or during storage. Under normal conditions, storage life will be 12 months before recharging.

2. Check battery charge. If not 1.270 specific gravity, charge batteries. See ELECTRICAL, Section 3.

3. Remove batteries from tractor and store in a dry weatherproof area.

REMOVAL FROM STORAGE

1. Remove protective covering from tractor tires and seals from air cleaner inlet, exhaust muffler, fuel tank breather and air conditioner air intake screens.

2. Remove blocks. Lower tractor onto tires.

3. Correct any leaks.

4. Inflate tires to recommended pressure.

5. Install fully charged batteries. Tighten cable clamps at both ends of cables.

6. Tension alternator, air conditioner compressor and fan belts.

storage

---

2. Subheads: STORING THE TRACTOR, STORING BATTERIES, REMOVAL FROM STORAGE

- Subhead type size somewhat smaller than general heading, but larger than the type size for subsystems that fall under each subhead.

- All uppercase letters reinforce hierarchical importance.

3. Subsystems: "Preparation," "Storing"

- Combination of uppercase and lowercase letters plus still smaller print size indicates lower hierarchical level.

4. Boldface: used for subheads and subsystems

- Reinforces distinction between headings and the text proper.

5. Boxes: used for safety warnings

- Boxes plus safety alert symbol in color call special attention to fire hazard.

Paragraph Clusters, Subheads, Core Sentences

Hierarchy of ideas can also be suggested by the way you handle paragraph clusters. For example, a mechanism or process may have a number of subsystems or subroutines that you want to treat as a unit in a paragraph cluster. Subheads and core sentences are useful devices to keep paragraph clusters focused and to organize and highlight key ideas of the paragraph(s). Core sentences serve as forecasting devices for what follows in the rest of the paragraph or section. Look at Example 3.1.

Comment   Example 3.1 shows how the initial (core) sentences set up expectations of subsequent development of the paragraph cluster. The word *three* in sentence 1 and the list of three steps in sentence 2 are then picked up and repeated, in the same order, in the subheads that follow. Commonsense principles of organization like these are simple to apply but easy to forget in the rush of putting together a manual. For the reader, they are the road signs that help the reader feel, "Okay, I'm still with you. I'm not getting lost."

**Example 3.1   Paragraph Cluster with Subheads and
Core Sentences**

OPERATION OF THE SPECTROPHOTOMETER

To operate the Ace Model X Spectrophotometer, three procedures must be followed. These include preliminary steps,
placement of samples in the cell compartment, and measurement of samples.

Preliminary Steps

Before making any measurements, turn the sensitivity switch
to standby, the shutter switch to SHTR, and the power and
lamp switches to the ON position. Then allow the machine to
warm up for 15 minutes before attempting any measurements. . . .

Placement of Samples in Cell Compartment

Each machine has two rectangular-shaped test tubes called
cells. One of these cells contains a reference sample. This
sample . . .

Measurement of the Samples

Slide the cell holder into position so that the reference cell is in
the light path of the compartment. Select the desired wave-
length . . .

Syntax Strategies

Elements in single sentences or groups of sentences offer additional opportunities for writers to provide road signs for the
reader. Among the most useful for manual writers are lists,
parallels, series, and comparison-contrast.

List Strategy

When steps or sequences are being described, the list strategy is
one of your most valuable tools. Many manuals rely too heavily
on the linear mode, stringing out a series of instructions hori-

zontally in long, complicated sentences and paragraphs. The horizontal arrangement makes instructions more difficult to follow. For example, if a novice photographer is developing film for the first time in a home darkroom, he or she will be following a series of steps very closely. Seeing those steps numbered and arranged vertically helps the user to keep his or her place. The user tracks the process by thinking, "Step 1 finished. Now for step 2."

The list strategy makes use of a powerful communication tool, the vertical. To understand how it works, add the following set: 456 + 1678 + 45 + 789 + 9. Keep track of how long it takes you to do this. Now add:

```
 357
4789
  23
 540
   8
────
```

The number of units, tens, hundreds, and thousands is the same in both sets, but most people are far quicker with the vertical. The list is the verbal equivalent of the numerical set. In Example 3.2 the words *before* and *when* serve as predictable organizers, like units and tens, and the reader's eye picks up only the key words as it sweeps down the passage. Note also that the white space around each element makes the list even clearer.

Parallels

When procedures or mechanisms are closely related, parallel sentence strategies also serve to sharpen focus. Notice the difference in clarity in the following examples:

Nonparallel

A. Install front bolts with the threads down. On the rear bolts, make sure the threads face up.

Parallel

B. Install front bolts with threads down.
   Install rear bolts with threads up.

Readers can understand instruction *A*, but they will grasp *B* more quickly.

Example 3.2.   Linear Mode Versus List Strategy

The following paragraph is written in *linear* mode:

The system must be vented under the following circumstances:
Before starting an engine that has not been operated for an extended period of time. When the fuel filters have been replaced. When an engine, in operation, runs out of fuel. When any connections between the injection pump and fuel tank have been loosened or broken for any reason.

Rewritten, its focus is sharpened by the *list:*

The system must be vented under the following circumstances:

—Before starting an engine that has not been operated for an extended period of time

—When fuel filters have been replaced

—When an engine, in operation, runs out of fuel

—When any connections between injection pump and fuel tank have been loosened or broken

Series

If you are listing a series of parts or steps, keep the series grammatically consistent:

Inconsistent verb forms

A. Always wait for the tractor to come to a complete stop. After lowering the equipment to the ground, make sure the transmission is shifted to the N position; the park brake should be set to prevent the tractor from rolling. Then remove the key.

Consistent verb forms

B. Always wait for the tractor to come to a complete stop, then lower equipment to the ground, shift the transmission to N position, set the park brake so the tractor will not roll, and remove key.

Readers will understand *A*, but they will have to slog through shifts from active to passive voice and unnecessary glitches in verb tense sequence ("wait," "after lowering," "is shifted," "should be set," "remove"). In *B*, each element of the series begins with an imperative verb ("wait," "lower," "shift," "set," "remove"). The series of commands is predictable, comfortable—in short, a clear map of what to do.

Comparison-Contrast

If you are trying to make comparisons or contrasts between steps or characteristics of a process or product, use parallel structure to heighten the comparison:

Nonparallel comparison:

A. More modern mechanisms have electronic controls, whereas they were formerly operated manually.

Parallel comparison:

B. Modern mechanisms are electronically controlled, whereas formerly mechanisms were manually controlled.

The first example is comprehensible, but the comparison is not sharp.

Sequencing

General to Specific

We have said that users respond to unfamiliar products and their manuals by searching for landmarks. Another way to think about this is to consider how you respond to meeting a stranger. You begin with a general impression of a person strange to you and only later begin to notice and understand details of dress, manner, or speech. The principle is a simple one—it is psychologically more natural to move from general to specific. Knowing this, you should try to introduce your user to the product by arranging the material of the manual from general to particular.

One way to assure this progression from general to specific is to think of your manual in terms of *overview* first, then *details*. Suppose, for example, that you want to provide instructions for a stereo system. The first section of the manual should contain an overview (a labeled, full-shot photo or drawing, a verbal description, or both) of the entire system: tuner, amplifier, tape deck, turntable, and speakers. This overview gives the reader a sense of how the parts relate to each other. Then, by means of details (subsystems), you may begin to provide more detailed information for each of the parts.

This same general-to-specific strategy also holds true for descriptions of processes or procedures. For example, before you begin to describe the individual chemicals and processes in a home permanent wave, tell the reader, by means of an overview, that the process will take well over an hour and include several subset processes: testing the hair for reaction to chemicals; preparing the hair (trimming, washing, etc.); applying curling lotion and curlers; rinsing and neutralizing; shampooing; setting the hair; and caring for the hair after the permanent. The overview can be brief—a list will do—but it sets up a forecast or expectation of the subroutines to follow.

Use the following guide as a way of ensuring a general-to-specific arrangement:

> Overview—introductory paragraphs or sections, with accompanying overview photo or drawing explaining the system, machine, process, or product
>
> Details—components of system, mechanism, or procedure

Spatial

For very large products, such as industrial machines, trucks, tractors, cranes, and motorcycles, users say they approach the product with a preconceived spatial logic. For example, they may think of the product from front to back, from top to bottom, or from left to right when facing the product. Of course, variations of this basic perception will be required in specialized manuals or service instructions. The logical view for service manuals on exhaust systems would be to show and describe the system as seen from below by the mechanic as he or she works with the system overhead on a hoist.

Whatever the product or process, it helps to ask your person-on-the-street user, "How do you think of this? Do you stand in front or at the side? Do you think of this from front to back? Top to bottom?" Then arrange the manual to match the user's spatial perception. Do not, for example, begin with the rear axle assembly on a truck or tell the user of a sewing machine how to embroider or buttonhole before you explain how to thread the machine or sew a straight seam.

Chronology

Most processes and procedures have an inherent chronology, that is, the steps for doing something grow naturally out of the way the product or process works. For example, the user will usually want to know about setup or assembly before operating procedures, maintenance, or storage.

However, at the level of paragraph or subsection, exceptions to strict chronology are quite common. Suppose that you are describing the operation and use of a home whirlpool. The dangers posed by high water temperatures to the elderly and to people with heart conditions or high blood pressure need to be mentioned, both in the manual and on the product, *before* the owner uses the whirlpool. Likewise, you would not instruct a user how to remove the cover of a pressure cooker without first showing how to make sure that all the steam has escaped from the cooker. In short, if any step or procedure can, in its execution, cause damage to the product or injury to the operator, be sure to explain this before the step is listed.

Anticipate trouble spots in procedures, even if chronology is interrupted. Do not warn of troubles or dangers when it is too late.

Patterns for Describing Mechanisms and Processes

Some writers find it useful to work from a rough organizational outline to ensure a general-to-specific arrangement of manual segments. Below are two outlines that you may use as guide-

Example 3.3.   Maintaining Proper Chronology

Finishing Your Home-Built Furniture

Advice too late

1. Prepare wood surface by using fine sandpaper or steel wool.
2. Mix oil-based stain of your choice.
3. Apply stain thinly with brush or cloth.
4. Allow to stand 5 minutes.
5. Remove stain by wiping.
6. Allow stained surface to dry for 24 hours.
7. Apply final finish of urethane or varnish.

Note: Some soft woods absorb stain quickly. If you are unsure of your wood, apply stain to test area first.

Advice on time

Many oil-based stains work best on hardwoods like cherry, walnut, or oak. If your furniture is made of soft wood, such as pine, test a sample area first to check absorbancy of stain. If stain soaks or smears, prime surface with thinned shellac before finishing. (Thinned shellac: 2 parts solvent/1 part shellac)

1. Prepare wood surface . . .
2. Test sample area for stain absorption.
3. Apply thinned shellac primer if necessary and allow to dry thoroughly.
4. Apply stain thinly . . .

lines. Be flexible in using them to make sure they match the special features of your product.

Pattern for a Description of a Mechanism

1. Introduction
   a. Definition
   b. Purpose

      c. General description

      d. Division of a device into its components

  2. Principle or theory of operation

      a. Divisions—what the part is, its purpose, and its appearance

      b. Divisions into subparts

        (i)   Purpose

        (ii) Appearance—often through visuals

        (iii) Details—shape, size, relationship to other parts, connections, material

  3. The operation of the system

      a. Ways in which each division achieves its purpose

      b. Causes and effects of the device in operation

Pattern for a Description of a Process

  1. Introduction

      a. General information as to why, where, when, by whom, and in what way the process is performed or occurs

      b. List of the main steps

      c. List of the components involved

  2. Description of the steps or analysis of the action

      a. First main step (or sequence of events)

        (i) Definition

        (ii) Special materials

      b. Division in substeps

  3. Conclusion (summary statement about the purpose, operation, and evaluation of the whole process)

Combining Strategies

We have described a number of writing and organizational strategies:

- General headings (sections or chapters)
- Paragraph clusters (with subheads and core sentences)
- Syntax strategies (lists, series, parallels, comparison-contrast)
- Sequencing strategies (general to specific, spatial, chronological)
- Patterns for describing mechanisms and processes)

These strategies seldom occur in isolation; rather, writers tend to combine the techniques to ensure smooth flow and internally consistent logic. Look at Figures 3.2, 3.3, 3.4, and 3.5 to see how writers have combined a number of strategies.

Figure 3.2.   Combination of Strategies: Highlighted Example to Show General-to-Specific Sequence, Parallels. This passage begins with a general description of valve location and function. Individual treatment of each valve is forecast with ''three or four'' in sentence 1 and reinforced with numerical opening in the paragraph cluster: inner or first, second, third, fourth. *Source: International 4386, 4586 and 4786 Tractors* 1096 026R2 5-78, International Harvester Company. Used by permission.

Original Text

Your tractor is equipped with three or four auxiliary valves. They are located on the right rear of the cab and are used to provide hydraulic control of various mounted and trailing-type equipment. The control levers are located to the right side of the operator on the cab wall.

Each valve provides a independent lifting and lowering operation. The valve may be set for float operation when equipment is to follow the ground contour. However, when the levers are operated at the same time or with the three-point hitch, the cylinder with the lightest load will move before the more heavily loaded one moves.

The inner or first valve lever operates the lower left rear hydraulic outlet. This lever also operates the 3 point hitch when tractor is so equipped.

The second valve lever operates the upper left rear hydraulic outlet.

The third valve lever operates the upper right rear hydraulic outlet.

The fourth valve lever operates the lower right rear hydraulic outlet.

(a)

Highlighted Version of Text

~~Your tractor is equipped with~~ (three or four) auxiliary valves. ~~They are~~ located on the right rear of the cab and are used to provide hydraulic control ~~of various mounted and trailing-type equipment. The~~ control levers are located to the right side of the operator on the cab wall.

~~Each~~ valve provides a independent lifting and lowering operation. ~~The valve may be set for float operation when equipment is to follow the ground contour. However, when the levers are operated at the same time or with the three point hitch, the cylinder with the lightest load will move before the more heavily loaded one moves.~~

1) The inner or first valve lever operates the lower left rear hydraulic outlet. This lever also operates the 3 point hitch when tractor is so equipped.

2) The second valve lever operates the upper left rear hydraulic outlet.

3) The third valve lever operates the upper right rear hydraulic outlet.

4) The fourth valve lever operates the lower right rear hydraulic outlet.

(b)

Figure 3.3. Combination of Strategies: General Heading, Subheads, Paragraph Cluster, Parallels, Spatial Sequence. The general heading, *Operation*, subsumes subheads including the three-paragraph cluster *Forward Movement, Reverse Movement,* and *Turning*. Within the paragraph clusters, individual paragraphs are further differentiated by *parallels*—to stop, to move, to turn; combinations of boldface, all uppercase, and uppercase and lowercase; *spatial sequence*—forward, backward, turning—right, left, circle turning. *Source: Ford 340 Compact Loader,* 1971, Ford Tractor Operations, Ford Motor Company. Used by permission.

OPERATION *General Heading*

foot throttle will not effectively aid in starting the engine. If necessary, use the choke control to assist in starting the engine.

3. Turn the key in the key switch (10), Figure 3, to its "ON" position.

4. Depress the starter button (9), to start the engine. Release the starter button when the engine has started.

IMPORTANT: *The starting motor must not be operated for more than a 30-second interval. If the engine fails to start within this period of time, wait 10 or 15 seconds before attempting again. If the engine does not start with a reasonable time refer to* STARTING DIFFICULTIES *in the Wisconsin Engine instruction book.*

NOTE: *During cold weather, use the choke control (5), Figure 3, to assist in starting the engine. After the engine starts, push the choke in. Allow the engine to warm up for one or two minutes before operating the unit.*

 CAUTION: *Never enter or leave the machine while the engine is running.*

Paragraph Clusters

STOPPING THE ENGINE

Prior to stopping the engine, lower the lift arms down against the frame and drop the bucket down to contact the ground. Turn the key in the key switch to its "OFF" position, to stop the engine.

NOTE: *If the unit is equipped with a parking brake lock, apply the Mico brake lock.*

FORWARD MOVEMENT

To move the loader forward, place the hands on both the left and right drive control levers (1) and (4), Figure 3. Ease both levers forward in unison and depress the foot throttle (12), Figure 3.

To stop the forward movement of the loader, ease both drive control levers backward in unison, thru neutral into the rearward position while releasing pressure on the foot throttle. When the forward movement stops, move the levers to neutral and depress the brake pedal to hold the loader in position.

NOTE: *Only three to five pounds force is required to move the control levers from neutral to full forward or full reverse. When the lever has reached the maximum of its movement, additional force will put undue strain on the valve spool and not increase performance. Best performance will result from easy handling and "feathering" the levers. Abrupt handling will not increase performance.*

REVERSE MOVEMENT

To move the loader rearward, place the hands on both the left and right drive control levers. Ease both levers rearward in unison and depress the foot throttle.

To stop the rearward movement of the loader, ease both drive control levers forward in unison thru neutral into the forward position while releasing pressure on the foot throttle. When the rearward movement stops, move the levers to neutral and depress the brake pedal to hold the loader in position.

TURNING

Turning Right: To turn the vehicle to the right, leave the right control lever in neutral and push the left lever forward.

Turning Left: To turn the vehicle to the left, leave the left control lever in the neutral position and push the right lever forward.

Tight Circle Turning: To turn the vehicle in a tight circle, either right or left, push one control lever completely forward and the other lever in full reverse position. The speed of the turn is controlled by the foot throttle pressure.

OPERATING THE LIFT ARMS AND DUMPING THE BUCKET

Raising and Lowering the Lift Arms

To raise the lift arms, move the lift control lever handle (2), Figure 3, on the right drive control lever to an upright position with a wrist action. To retain the arms in any position, release pressure and allow the handle to return to the neutral position. To lower the load arms, twist the handle down to a horizontal position. To stop the descent, allow the handle to return to neutral.

Figure 3.4.   Combination of Strategies: General to Specific, Paragraph Clusters. General to specific is shown by the descending print size of headings. Notice how the paragraph clusters are sharpened by boldface, white space, and the dictionary or glossary format. These strategies separate discrete features and functions of system parts. *Source: International 596 Tandem Disk Harrow*, 1096 230R1 3-79, International Harvester Company. Used by permission.

WORK SAFELY — FOLLOW THESE RULES

Hydraulic fluid escaping under pressure can have enough force to penetrate the skin. Hydraulic fluid may also infect a minor cut or opening in the skin. **If injured by escaping fluid, see a doctor at once.** Serious infection or reaction can result if medical treatment is not given immediately. Make sure all connections are tight and that hoses and lines are in good condition before applying pressure to the system. Relieve all pressure before disconnecting the lines or performing other work on the hydraulic system. To find a leak under pressure use a small piece of cardboard or wood. Never use hands.

INTRODUCTION

HARROW FEATURES

Hitch coupling tongue has large bearing area to reduce wobble and wear.

Spring-cushioned self-leveling hitch protects the disk blades and the entire harrow by reducing shock loads. The self-leveling feature assures uniform penetration.

Unit-built gangs have big 1-1/2-inch (38 mm) spring steel arbor bolts and heavy welded standards with a drop-out bearing design.

Hydraulic Control allows the operator to fold or unfold the wings, raise or lower the harrow, and set the gangs to a precise depth for working.

Heavy, all welded main frame and wing frame of 8 x 6-inch and 6 x 4-inch rectangular tubing and gang bars of 10 x 4-inch rectangular tubing.

Large flange extra heavy duty spacing and bearing spools have machined ends to assure precise contact with the disk blades and absolute squareness of the entire gang.

Disk blades are crimped center design of heavy gauge steel to reduce blade breakage. The blade edges retain their sharpness longer, resulting in a better disking job. To provide a furrow free leveling action, a disk blade 2-inches (51 mm) smaller in diameter than the basic blade size is used on the outer ends of the front gangs and a two blade taper is used on the outer ends of the rear gangs.

Gang bearings are regreasable cartridge-type triple-sealed, self-aligning ball bearings.

Center tooth is used to eliminate the uncut material between the front gangs.

Dual wheels carry the harrow for transporting and act as gauge wheels for precise depths. Wheels on the wings provide flotation and depth control over the entire harrow width. Dual wheels for wings are optional for 22 ft. 7-inch (6.87 m) harrow size, and standard on other sizes.

Scrapers are flexible spring steel type and are individually adjustable. Their high clearance design keeps the disk gang free from plugging as well as the blade surface clean. Scrapers mounted on the bearing standards provide optimum cleaning of this critical area.

Transport locks, consist of gauging straps to secure the disk gangs in a raised position for safe harrow transport or removing the cylinder. The gauging straps have a series of holes for depth control.

Hydraulic equipment required for transporting, wing folding, and depth control includes two master cylinders, two slave cylinders, two wing fold cylinders, hoses, hose clamps, tie straps, and connections. These are furnished with the harrow and are for tractors with two hydraulic valves.

 CAUTION! Both transport pins must be used; one in each pair of gauging bars when transporting the harrow.

Figure 3.5.   Combination of Strategies: General Headings, Subheads, and Core Sentences, General-to-Specific Sequence. This three-page section of a manual combines several strategies: 1. Headings and subheads—correspond to facing page, full-shot (labeled) view of the product. Photo and opening paragraph give an overview and introduction to the product; 2. Core sentences—forecast subsequent detailed treatment of systems; 3. General to specific—section begins with a general introduction to the product and its uses: the first five elements (Controls, Hydrostatic Drive, Hydraulic System, Steering, Wheels and Tires) are briefly described; then fuller treatment of each of the five elements follows, beginning with Controls. The shift to fuller treatment is indicated by a move to larger print. *Source: Ford 340 Compact Loader,* 1971, Ford Tractor Operations, Ford Motor Company. Used by permission.

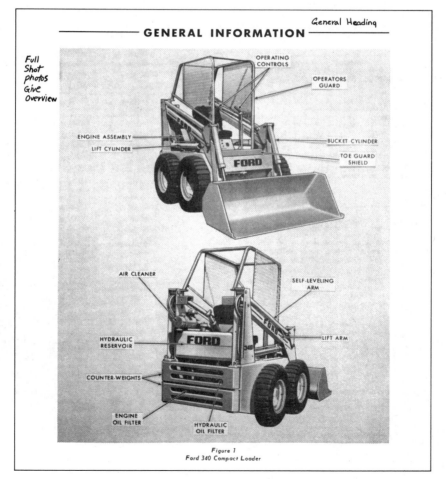

Figure 3.5. *(continued)*

General Heading (repeated)

GENERAL INFORMATION

INTRODUCTION

↓ Introductory Statement: Overview

The ~~Ford Series 340~~ Compact Loader, Figure 1, is designed for handling construction, industrial, mining, and agricultural materials. In its capacity as a self-contained rider-type mechanized piece of handling equipment, the loader is completely equipped to lift loads of a 1500 pound capacity to a maximum height of 108 inches at the tool bar hinge pin. The vehicle is designed for transporting of loads from one area to another, both indoors and out-of-doors. With the variety of attachments available, the ~~Ford Series 340~~ Compact Loader can be equipped to perform a multitude of job site uses.

The following paragraphs contain a description of the components and systems incorporated in the loader.

ENGINE ASSEMBLY *(Treated Elsewhere)*

The engine assembly, Figure 1, is located directly behind the operator's seat. The engine is a four cylinder, L-head, air cooled, 30-horsepower, Model ~~VH4D,~~ Wisconsin gasoline engine. <u>Refer to the engine instruction book</u> for further information concerning the engine; its components, specifications, and repair. Standard engine accessories include a dry-type air cleaner, oil filter, a 12-volt dust-sealed alternator, starter, fuel pump, velocity governor, and two series-connected five gallon fuel tanks.

Identical Type Face (1-5)

1. OPERATING <u>CONTROLS</u> AND <u>INSTRUMENTS</u>

Standard <u>instruments</u> are a key ignition switch, safety starter button, ammeter, and hourmeter. Con-trols consist of the choke, foot throttle, brake pedal, and the hydraulic lift, bucket, and drive controls.

List of components ↑↑

2. HYDROSTATIC DRIVE

Engine-driven hydraulic pumps provide four-wheel drive with separate power circuits for each side of the vehicle. The drive train is sealed in oil inside dust-proof cases, providing maximum protection.

3. HYDRAULIC SYSTEM

The unit contains a separate hydraulic lifting circuit. Two, 2-1/2 inch diameter double-acting cylinders provide lift and lower to the boom arm. Two 2-1/2 inch diameter double-acting bucket cylinders provide downward movement, or tilt, to the bucket or attachment. The system has a 14-gallon oil capacity, filters, filter gauge, oil cooler, and check valve.

4. STEERING

Steering is accomplished by a valving arrangement which controls the hydraulic drive circuit to each side of the vehicle. Difference in hydraulic flow drives one side faster or slower than the other side and results in turning. One side full forward and the other side full reverse gives a full pivot turn.

5. WHEELS AND TIRES

The unit has heavy duty 15-inch wheels, each separately mounted on heavy duty axles and 7:50 x 15, 6-ply special loader tread pneumatic tires all around. Hydraulic brakes are provided on the front wheels.

FRAME ASSEMBLY

The frame assembly is built of heavy steel electrically welded into an integral unit. Steel re-enforcing creates maximum resistance to stress and strain.

Shift to larger type — Fuller Treatment of #1

1. CONTROLS AND INSTRUMENTS

Before operating your new ~~Ford Series 340~~ Compact Loader, thoroughly familiarize yourself with the location and function of all <u>controls</u> and <u>instruments</u>.

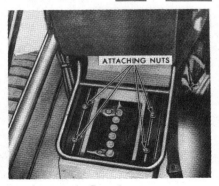

ATTACHING NUTS

Figure 2
Seat Adjustment (Cushion Removed)
Anticipates Next Page

Figure 3.5. *(continued)*

CONTROLS ——— *New General Heading*

SEAT AND SEAT ADJUSTMENT

The ~~Ford Series 340~~ Compact Loader is equipped with a contoured-type seat. The cushion portion of the seat is removable to gain access to the battery. If the seat is not in the most comfortable position, it can be adjusted fore and aft by loosening and attaching nuts, Figure 2, and repositioning the seat as desired. After the seat has been adjusted, tighten the attaching nuts securely.

SEAT BELT
General
The seat belt, Figure 3, is located and installed as are conventional automobile seat belts. Always fasten the seat belt whenever seated in the operator's seat.
Particular
To lengthen the belt, tip the buckle end down, then pull on the buckle until the ends can be joined.

To shorten the belt, buckle it, then pull on the loose end until the belt is snug.

Figure 3
Controls and Instruments

1. Right Drive Control Lever
2. Lift Control Lever
3. Bucket Control Lever
4. Left Drive Control Lever
5. Choke Control

6. Brake Pedal
7. Auxiliary Hydraulic Treadle (Optional)
8. Engine Hourmeter
9. Starter Button
10. Key Switch

11. Ammeter
12. Foot Throttle
13. Hand Throttle (Optional)
14. Seat Belt

Strategies for Reducing the Length of Verbal Text

Manuals are seldom read under the relaxed conditions in which one might, for example, read a novel or a technical reference book. Because manual users are reading and dealing with the product at the same time (starting, adjusting, repairing), you should try to balance sections of solid prose, arranged in paragraphs, with other information-providing strategies. A number of writing strategies can be used to reduce the bulk of solid prose text. The most obvious of these are visuals (Chapter 5) and formatting (Chapter 4). Other techniques that serve to provide relief from too much print and to make manuals more quickly comprehensible are

- Charts and tables
- Illustrations and examples
- Omission of unnecessary theoretical/technical background

Charts and Tables

Whenever you have numerical information to present or when you have to relate items of information (for example, operating conditions and frequency of maintenance), use a chart or table. Readers find it very difficult to keep track of numbers buried in paragraphs of prose. It is much better to pull the numbers together in a table and then discuss them (with reference to the table) in a paragraph or two.

When you set up a table or chart, keep these principles in mind:

1. Arrange the headings and data in a rational order.
2. Vertically arrange the items that are to be directly compared. It is much harder to compare items horizontally. Compare: $35.64, $27.83, $56.27 and

   $35.64
   $27.83
   $56.27

3. Include units of measurement in all headings.
4. Align columns of numbers on the decimal point.
5. Use lines to divide columns and rows only if confusion is

likely without them. Lines add to clutter. If possible, use white space as a divider instead. (See Figure 3.6.)

Figure 3.6. Two Versions of the Same Chart—One With Lines Dividing the Columns and One Without. Note how the absence of the lines in the second version reduces visual clutter. Note also, however, that the rectangular shape of the white space takes the place of a line.

LINE CAPACITY OF SPOOL (IN YARDS)		
Lb. Test	Large	Small
2	---	300
4	400	250
6	350	200
8	300	150
10	250	100
12	200	---
15	150	---
18	100	---

LINE CAPACITY OF SPOOL (IN YARDS)		
Lb. Test	Large	Small
2	---	300
4	400	250
6	350	200
8	300	150
10	250	100
12	200	---
15	150	---
18	100	---

Figure 3.7. Two Versions of Another Chart—Again, With and Without Lines Dividing the Columns. Note here that the lines (or wider spacing of the columns) are essential to avoid confusion because the shape of the white space is so irregular.

PARTS LIST (PARTIAL)			
Part Name	Order No.	Part No.	Price
Axle	305	81-214	$2.25
Baffle plate	6342-R	56	1.20
Click spring	11	322-D	.30
Drive gear	452-T	9120	4.50
Pivot	6783-DE	3	.75
Rotating head	66	81-615-A	11.45
Spool	43598-OS2	27	3.75
Transfer gear	452 (453)	16 (17)	.75 (.85)
Trip lever	340972	81-005	1.25

PARTS LIST (PARTIAL)			
Part Name	Order No.	Part No.	Price
Axle	305	81-214	$2.25
Baffle plate	6342-R	56	1.20
Click spring	11	322-D	.30
Drive gear	452-T	9120	4.50
Pivot	6783-DE	3	.75
Rotating head	66	81-615-A	11.45
Spool	43598-OS2	27	3.75
Transfer gear	452 (453)	16 (17)	.75 (.85)
Trip lever	340972	81-005	1.25

Figure 3.8.   Two-Color Chart Used to Divide Columns and Rows. Gray and white used for column and row dividers. Note how numerical information has been handled as a unit (5 hours, 10 hours, etc.) to prevent unnecessary repetition of numbers. *Source: 2200 Seed Drill, Operator's/Set Up/Parts Manual,* 1983, Versatile Farm Equipment Company, a division of Versatile Corporation. Used by permission.

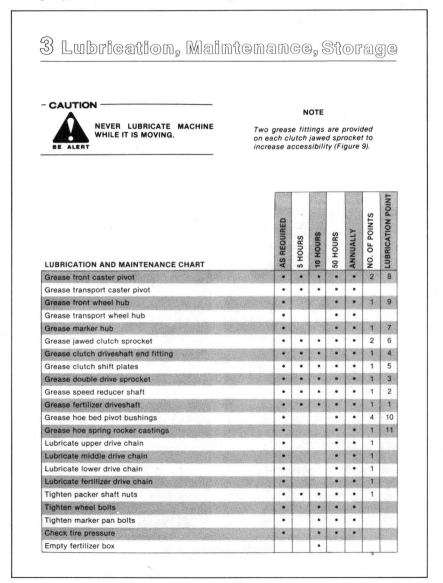

To work well as a divider, the white space must be regular in shape and more or less rectangular. (See Figure 3.7.) Color may also be used to divide columns and rows. (See Figure 3.8.)

Illustrations and Examples

Illustrations and examples provide another way to reduce the length of written text. They can show alternative ways of carrying out a process or can distinguish between the right and wrong ways of doing things. If you look back at the strategies suggested in this chapter, you will see that they are supported by illustrations and examples. For instance, after you read a description of the list strategy, you were provided with an example showing conversion of a linear passage to a list.

Illustrations and examples may also be used in combination with written text as a shorthand to describe a procedure more briefly. In essence, what you are doing is showing users, rather than merely describing in words, how a product will look after a step or procedure has been completed. Illustrations and examples are especially useful for products whose operation involves many steps or procedures. Manuals for electronic devices, monitors, and computers used for graphics, numbers or word processing all make use of examples and illustrations to clarify what the written text describes. (See Figure 3.9.)

Omission of Unnecessary Theoretical/Technical Background

Good organization of a manual means knowing not only what to include, but also what to leave out. Manual writers with technical backgrounds are often tempted to include too much theoretical material. The theory or experimental design that underlies a product will always be of interest to the manufacturer as well as to the technically sophisticated manual user. However, in manuals for the general public, you should be guided most of all by what the user needs to know.

Not only does excessive theoretical explanation take up page space and increase manual costs, it is also forbidding and distracting for naive manual users.

Figure 3.9. **Examples Used to Demonstrate What Written Text Is Describing.** This manual passage shows how instructions for designing a computer program consist of two or three sentences of written instructions followed by an example of how the program entry will look. *Source: Micro-mode Mongean Descriptive Geometry* by James McNeary and Robert Perras, 1982, unpublished. Used by permission.

```
Now is a good time to store the constant values in the
computer.  We want to enter our aspect ratio scaling factor
AR, and our metric to computer units conversion factor MC.        } Text
We also want to turn on and clear the graphics screen, set
the color of the display to white, and set the scale of our
shape table drawings to one:

     130 HGR:  HCOLOR = 3:  SCALE=1:  AR=1.18:  MC=2.44        Example

We want to use the entire graphics screen area, so we turn    } Text
off the text display normally shown on the bottom four
lines:

     140 POKE -16302,0                                           Example

The location of point B has been entered into the computer
at line 20.  Now we must take those metric values, and
convert them into computer units before we can draw the
points on the screen.  To do this, we set a new value for
the point equal to the old value multiplied by our           } Text
metric/computer units scaling factor (MC).  Make note of
the form used to do this.  The new value is always on the
left side of the equation.  The old value of the same
variable is used in the calculation on the right side of
the equal sign:

     150 X(1)=X(1)*MC:  Y(1)=Y(1)*MC:  Z(1)=Z(1)*MC          Example

Now we should position the principle fold line on the
screen.  It will be drawn from a starting point we will
designate F1,H2.  F1 represents the X value on the monitor   } Text
screen; H2 represents the Y value.  From our shape table
matrix (Figure 8), we know that the fold line drawing is
shape #87:

     160 F1=40:  H2=80:  DRAW 87 AT F1,H2                     Example

Now we must determine mathematically where point B is to be
located in relation to the principle fold line just
plotted.  Remember that X is the distance of the point from
the imaginary reference line (Figure 10).  Therefore, for
the point to be properly located on the screen, we must add
the "X" value of the principle fold line starting point
(F1) to the X value of point B.

Y is the transfer distance of point B back from the "f1,h2"
reference line.  To locate the point on the monitor in the
proper position on the Horizontal Projection Plane, which
is above this principle fold line, we must SUBTRACT the
value for Y from the "Y" value we gave to the fold line
(h2).

Z is the vertical transfer distance of point B from the
principle fold line.  To position the point properly on the
monitor in the Frontal Projection Plane, we must ADD the
value of Z to the "Y" value we gave to the fold line (h2).
(i.e. remember the monitor's coordinate system.)
```

Some samples of excessive theoretical/technical material:

- A manual for a surface tension measurement device in which the first four pages describe the history of surface tension study, going all the way back to Archimedes.
- A manual for an inexpensive stereo system that describes several decades of development work on printed circuits.
- A manual for a sailboat that devotes over half its space to unresolved controversies on vector theory and wind motion.

You must use your judgment to determine how much theory is appropriate for adequate understanding of your product, but remember that the average buyer of a microwave oven, for example, will probably not be interested in the esoteric physics of microwaves. Much of the material found in operator manuals for the general public would be more appropriate in the service manual. (See Chapter 7.)

Review of Effective Writing Strategies

Using Example 3.4, take a minute to review your understanding of the writing strategies covered in this chapter.

Comment The two sets of instructions for the tape recorder–radio have the same information, but *B* is more effective than *A*. Passage *B* is easier to follow because

- Sentences are shorter and begin with active verbs.
- The variety of typefaces and use of boldface for names of controls helps the user locate controls more quickly on the tape recorder itself. The writer uses caps for all buttons (STOP, PLAY, EJECT, OFF) and boldface and initial caps for all parts of the player (Cassette Door).
- Wording for controls exactly matches wording on the product.
- Numbered-list strategy is used by both writers, but list effectiveness in Passage *A* is blurred by muddy, strung-out sentence patterns.

Example 3.4. Review of Effective Writing Strategies: Comparison of Two Sets of Instructions

Read these two sets of instructions for using a portable tape recorder–radio. How would you answer these questions?

1. Which set of instructions is easier to follow?
2. What writing strategies make that set easier?

Passage A

1. Before playing tapes, make sure the radio switch is turned off.

2. To open the cassette door, press the button that stops the player and ejects the tape.

3. Insert a cassette into position so that the tape side is facing upward and the tape itself is on the right. Then press the cassette into place securely.

4. Push the cassette door closed and press the play button.

5. If you want to adjust the sound, adjust the controls for volume level and tone.

Passage B

1. Set RADIO switch to OFF.

2. Press STOP/EJECT Button to open the Cassette Door.

3. Insert a Cassette into position as shown—open tape side up and full reel toward the right. Press the back of the cassette all the way in.

4. Close the Cassette Door and press PLAY.

5. Adjust VOLUME and TONE for desired sound.

Summary

Intelligent use of organization and writing strategies

- Makes the structure of the manual "jump" off the page
- Provides important landmarks for the reader
- Reduces the bulk of the verbal text

Pay special attention to headings, core sentences, paragraph clusters, and syntax strategies, and organize your material from general to specific. Be careful in deciding how much material to include, and minimize solid blocks of prose by the use of charts, tables, examples, and illustrations.

Tell users what they need to know.

Omit material that is merely nice to know.

4

Format and Mechanics

Overview

In the various chapters of this book, we discuss what kind of information and what kinds of supporting materials (photos, diagrams) to include in an instruction manual. In addition, we discuss how best to organize and write the manual. This chapter discusses how to arrange this material so that readers can easily find their way through it to the particular information they need. The best writing in the world is wasted if the pages are visually unappealing (users won't read them), the referencing is inadequate (they won't be able to find what they need), or the mechanical elements—paper, binder, etc.—are inadequate (the manual will be too hard to use). Skillful design choices in these matters, however, will maximize the manual's usefulness and help ensure that the user will read and refer to it often.

Format/Layout

Strictly speaking, *format* refers to the mechanical specifications of the page—page size, column width, type size, etc.—and *layout* refers to the placement of actual text (copy) and visuals (art) on the page. In this book, since we are giving general guidelines, we will use *format* to cover both areas—both have to do with the visual impression a page gives. This visual impression is remarkably important. All of us have picked up a manual or report, glanced through it, and thought, ''This looks too hard to bother with'' before we had actually read any of it. On the other hand, we have also picked up articles or reports and thought, ''This looks interesting. I think I'll read it''—again, before we had actually read any of it. The difference in the two responses is largely a result of the visual appeal of the page. In the first case, perhaps the type was too small to read easily or was set in one unbroken block, with no white space for eye relief. In the second case, perhaps eye-catching headings provided us with a sense of what was covered before we started reading—giving us a head start on understanding.

The writers of instruction manuals, more than many kinds of technical writers, must be conscious of this visual appeal. Managers will read technical reports because they must—they need the information. They may be frustrated by bad format, but they will still make an attempt to plow through the document. The user of a product, however, is all too likely to ignore the manual if it looks hard to use—ignore it, that is, until something breaks because he or she did not read the instructions and did not use the product properly.

You have four elements to work with in designing an effective format: the shape of the text, type size and style, headings, and white space. We will look at each of these elements separately, although they must work in concert for the format to be effective.

Shape of the Text

Look at a page, and draw an imaginary line around a block of text. This is what we mean by the shape of the text. It can be perfectly rectangular or irregular along the right-hand side; it can fill up the whole page or be one of several blocks on that page. In typesetting terms, the shape of the text is determined by column width, column depth, and the contour of the right-hand margin—that is, whether the margins are *justified* (even, like a newspaper column) or *ragged right* (like the right-hand margin of a typed page). If you work for a large company, the decisions about column width and so forth may already be determined for you. Most companies require that all their manuals follow a standard format. If you work for a small company, or if your company is just getting into the business of operator manuals, you may be called upon to make these decisions yourself. (See also Chapter 1, ''Planning.'') Here are some general guidelines to help you. (Even if you work in a company that requires a standard format, you may be able to improve on it.)

Column Width   Column width is a function of line length, which in turn is a function of the type you use. Readability studies have shown that the optimum line length is about an alphabet and a half, or roughly 40 characters:

abcdefghijklmnopqrstuvwxyzabcdefghijklm

With a line much shorter than that, the eye must keep jumping to the next line, and words and phrases are frequently broken, causing fatigue and decreased comprehension. If the line is much longer, the eye has too far to travel back to begin the next line and is likely to settle on the wrong line of type. (A corollary is that the longer the line, the more space is needed between lines.) Clearly, the smaller the type you use, the shorter will be a line composed of 40 characters. A normal typed page has a line length at least half again as long as the recommended length, but we are so used to it that readability seems not to be affected. However, in operator manuals, where the reader is likely to be looking at the manual, then at the product (to locate a part or do a procedure), then back at the manual, a two-column format facilitates the reader's finding his or her place easily. Remember that with a two-column format you can still run a photo or diagram the full width of the page if a one-column width would make it too small to read easily.

Column Length   Columns ordinarily run the full length of the page, although column length may be affected somewhat by the illustrations you use and by the way you use white space. (The use of white space is discussed in a subsequent section.)

Justified or Ragged Margins   In typeset material it is possible to justify the right-hand margin, that is, to make it even. This requires adjusting the spacing between words and letters to make all the lines come out the same length. It also ordinarily requires frequent hyphenation of words. There is no right or wrong decision about whether to justify the right-hand margin. Generally, justified copy is perceived as more formal and may be slightly harder to read.

Type Size and Style   All the various type styles may be divided into two groups: serif type and sans serif type. Serifs are the little lines (sometimes only suggested) at the ends of each stroke in a letter. Serif type has these little lines, and sans serif does not (*sans* is French for "without"). Figure 4.1 shows examples of serif and sans serif type.

Figure 4.1.   Samples of Serif (right) and Sans Serif (left) Type. Sans serif typefaces normally use strokes of a single weight, or width. The Omega typeface shown here is unusual, in that the lines widen out toward the ends—almost suggesting the beginnings of serifs.

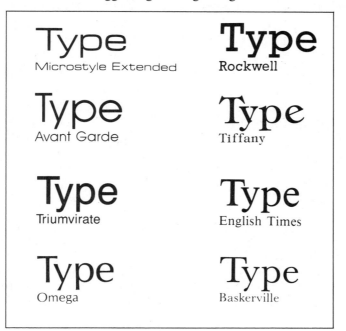

Sans serif type has a "modern" look to it, but readability studies indicate that most people perceive it as harder to read. No one is quite sure why this is true—it may be that the serifs tend to "seat" the letters on the line and pull the eye along to the next word. Sans serif type is fine for headings, but for the text proper we recommend using a serif type.

Similarly, a mixture of uppercase and lowercase letters is easier to read than text written in all uppercase. The reason for this is clear: lowercase letters show much more variation in shape than uppercase letters, permitting easier recognition.

Type size is measured in points, with 72 points to an inch. To understand what is meant when we say that something is set in, say, 12-point type, imagine a standard typewriter key. A little metal block with a raised letter on it hits the ribbon to make a letter appear on the page. In traditional typesetting, a page would be set by lining up a series of individual blocks with raised letters on them and locking these into a wooden frame to hold them in position. The point size of the type measures the height of the *block*—not just the letter itself. Thus, the letters in two different styles of 12-point type may be slightly different heights, but the blocks would be the same. Even though most typesetting today is done by computer, the old terms are still used.

You can see that the tallest letters in 12-point type would be a bit less than one-sixth of an inch high. Figure 4.2 shows examples of several type sizes. Most text is set in 8- to 12-point type. The smallest size that can be read without a magnifying glass is 6-point. Because operator manuals are likely to be read under less than ideal conditions, we would recommend going no smaller than 8-point type for anything in the manual and would urge using 10-point or 12-point for ease of reading.

Headings

Headings are signposts that give the reader a sense of what is covered in a section of the manual. (See also Chapter 3, "Organizing and Writing Strategies.") Most readers will skim over a section by reading just the headings before starting to read the details. For this reason you should make sure to

6 pt. Type size can affect readability.

8 pt. Type size can affect readability.

10 pt. Type size can affect readability.

12 pt. Type size can affect readability.

14 pt. Type size can affect readability.

16 pt. Type size can affect readability.

18 pt. Type size can affect readability.

20 pt. Type size can affect readability.

22 pt. Type size can affect readability.

24 pt. Type size can affect readability.

26 pt. Type size can affect readability.

28 pt. Type size can affect readabilit

30 pt. Type size can affect readab

32 pt. Type size can affect reade

34 pt. Type size can affect reac

36 pt. Type size can affect re

Figure 4.2. Samples of Different Type Sizes, Measured in Points. We do not recommend type smaller than 10- or 12-point for use in manuals.

- Use enough headings
- Use headings that reveal the organization of the section

Most writers use too few headings. Instead of just labeling the major sections of a chapter, consider adding subheadings to point out the smaller divisions. Remember that, as the writer, you are familiar with what is contained in the manual you are writing—you know where the parts are and what is covered in each. Your readers are looking at the manual for the first time; without sufficient headings to help them find their way, the manual will look like a bewildering sea of prose.

Be sure that your headings accurately reflect the organization of material in the manual. Use varying type sizes and boldface type to set the headings apart from the text and to give the reader clues about the structure. Thus, major divisions should be signalled by larger, boldface type, and minor divisions by correspondingly smaller type headings. Taken as a whole, the headings should form the skeleton, or outline, of the chapter. Not only will this technique help readers to understand the structure of the material, but it will also make it much easier for them to find a particular piece of information just by leafing through the book.

White Space

The judicious use of white space in an operator manual can improve both readability and comprehension and can provide your reader with another set of visual cues to organization. In times when the cost of paper is constantly rising, many writers are tempted to cram as much as possible onto each page in order to cut production cost. But this practice is akin to cramming everything possible into one photograph or diagram—you end up with a confusing jumble that the reader won't even bother trying to decipher. We have all found ourselves put off by one page of text and attracted by another, largely as a result of an unconscious (or conscious!) assessment of how difficult it would be to read. An important factor in this judgment is white space.

Adequate white space can *by itself* improve readability simply because it makes it physically easier to read the page. Our eyes, like the rest of our bodies, need rest breaks. If these mini rests are not provided to us in the form of white space, our eyes will take them anyway—and we will find ourselves having to reread a sentence because we missed a few words. Also, most people are able to read, that is, move their eyes along the line of text, faster than their minds can follow. White space gives the mind time to assimilate information before going on to the next piece.

We do not read word by word, piling individual words one on top of the other, but rather in chunks—adding phrases and sometimes whole sentences together to form the thought. For the mind to grasp a chunk of information, it must see it as a chunk, and white space can help with this. Even such a simple use as leaving a space between paragraphs makes a difference. It helps the mind to see the information in that paragraph as a unit.

White space can also give information about the structure of a piece of writing. In the same way that headings can show how the manual is organized, white space, if used carefully and consistently, can show how a section of text is put together. Thus, a blank line or two (such as between paragraphs) lets your reader know that you are moving from one unit to another of equal importance. Similarly, indenting a section (widening the white space around it) lets your reader know that you are moving to a smaller organizational division within a single unit. Note that surrounding an item with white space will also call attention to it—as in the alphabet-and-a-half column width example used earlier in this chapter.

Using white space to signal the organization of a chapter will work only if you use it consistently: if, for example, you always use the same number of blank lines between major divisions and always use the same (but smaller) number between minor divisions. Again, some of these decisions may already be made for you by company policy, but you may wish to go further than the minimum specifications of a standard format. You will find it easiest to be consistent if you make these decisions at the outline stage and make a little table to help you remember what you decided. Otherwise, you will be having to page back through completed material to find whether to leave two lines or three between sections.

Careful use of white space may add a tiny bit to the cost of a manual because not every page is crammed corner to corner, but it will help ensure that the manual is used. If the manual sits untouched on the shelf, the whole cost of producing it is wasted.

Referencing

Because people rarely read operator manuals cover to cover like a novel, referencing is extremely important to help the reader find the section he or she needs. You have three types of referencing available for use in writing an operator manual: the table of contents, the index, and cross-referencing. Let's look at these in turn.

Table of Contents

The table of contents is essentially a map of the book. It appears in the front of the manual and outlines what the manual covers, giving page numbers of the beginning of each section. The reader will probably look there first for a specific item of information. Any manual more than a few pages long should include a table of contents.

To be useful, the table of contents must be arranged so that it clearly shows the organization of the manual and is easy to use. The names used to designate sections in the table of contents should match the headings used in the text. You may not want to include every tiny subhead in the table of contents, but those you do include should have the same wording in both places. You should also show organizational levels in the table of contents, perhaps by type size or by indentation, to reflect the arrangement of information in the text. Be careful to make it readable. If the page numbers on the right are too far away from the words on the left, the reader may find it difficult to know what goes with what. You may wish to run dotted lines across the page or leave spaces between small groups of headings and page numbers. You must strike a balance between making the table of contents too skimpy and making it so complete that it nearly reproduces all the information in the manual itself.

In addition to (or instead of) the traditional table of contents, other means may be used to help a reader find his or her place. Some manuals use a tab system, whereby the sections of the manual are divided by tab pages to make it easy to flip to the section you need. Some use colored pages to distinguish one section from another. Some manuals use a two-level table of contents: a general one at the beginning and a specific one at the head of each section. Whatever method or combination of methods you choose, the same basic rules apply: make it logical, reflective of the organization of the manual itself, and easy to use.

Index

The index is an alphabetical listing of subjects and the numbers of the pages on which they appear. It is usually placed at the back of the manual and is more comprehensive and detailed than the table of contents. It does not, of course, show the structure of the manual, but it is useful for locating a specific item of information quickly. Preparation of an index used to be a tedious process of putting each entry on a separate index card and going through the text writing down page numbers as each entry appeared. Now, with the use of computer text editors and word processors, it is a simple matter of having the computer do the searching.

Cross-referencing

Often a reader using an operator manual needs to be aware of information in a section other than the one he or she is using at the moment. For example, in a lawn mower manual, the section on winter storage of the mower might say to drain the oil and replace it before the next use. The location of the drain plug and the proper weight oil to use is listed in the section on maintenance. Rather than let the reader page through the manual at random looking for this information, point it out: "See 'Maintenance,' page 4." Put yourself once again in the reader's shoes, and remember that he or she does not know the contents of the manual backward and forward as you do. Whenever you think it would be helpful for the reader to be referred to another section of the manual, do so. Of course, the most obvious place for cross-referencing is in the troubleshooting section. This is usually set up as a table with headings like "Problem," "Probable Cause," and "Remedy." Too often, the remedies suggest something like "adjust spark plug gap" without telling the reader where in the manual the proper gap is given. To save your readers a lot of frustration, always include the full information needed—title of section *and* page number. If they are using the troubleshooting section, they are frustrated enough already.

Mechanical Elements

The mechanical elements of a manual include the paper it is printed on, the cover, the binding, and the size. You must choose these as carefully as you choose the photos to be included or the manual's organizational structure. These seemingly superficial elements may make or break a manual because they affect ease of use and durability. No one set of rules will cover all applications, but in this section we will present some factors to consider as you make your choices.

Paper

Two major questions arise about the paper used in manuals:

1. How durable is it? Is the paper easily torn? Is it flimsy and likely to shred with hard use? Clearly, if your manual will be referred to again and again, you must use durable paper. In addition, thin, flimsy paper may allow bleed-through, that is, allow the print on one side of the page to be seen from the other side, which makes for harder reading.

2. How porous is it? Porosity will affect how the paper accepts ink and thus how sharply photos will reproduce. Porous paper will allow the ink to bleed, thus obscuring fine detail. Porous paper will also accept other substances, like oil and dirt. If your manual is likely to be used under dirty conditions, you should choose a harder-surfaced paper, possibly even a coated stock (although this is quite expensive).

Cover

The cover is both a mechanical element in the manual and a public relations device. Most companies have a standard cover format for their manuals, including information about the model and often a picture. This dual purpose dictates the questions you must ask yourself when you design a cover and choose the cover material.

1. Is it attractive? Will the user want to read it, and does it speak well for the company?
2. Is it durable? The manual cover protects the inside pages and must thus be made from heavier stock. If the manual is likely to be used in a harsh environment (rain, oil, etc.), it should be made of coated stock, possibly even of vinyl.

Binding

The purpose of the binding is to hold the pages together so that they can be easily read. Here are some factors to consider:

1. Will the pages lie flat? Nothing is more irritating than trying to do a procedure requiring both hands and frequent glances at the instructions, only to find that the manual flops shut each time you let go.
2. Will the pages begin to fall out after hard use? This is a common problem with "perfect" bindings, though rarely with stapled, stitched, or spiral bindings.
3. Will frequent additions or corrections be sent to owners? If the manual is likely to be updated often, it might be a good idea to put it in a three-ring binder, so that outdated pages can be easily replaced with new ones. If the binding requires that holes be punched in the pages, make sure that margins are wide enough to allow this to happen without losing part of the text.

Size

The size of the manual is related to how it will be used. An 8½-by 11-inch three-ring binder would be an awkward size for a manual for a 35 mm camera—you want something small to fit in a camera bag. On the other hand, 8½ by 11 is a good size to use on a workbench in a garage. Think about how your customer will want to use the manual. Will he or she want to tuck it in a pocket or have it handy on a shelf? In general, small, oddly shaped manuals are easier to lose than more standard-size ones, but there are no hard-and-fast rules. As always, a clear understanding of your audience and of the manual's purpose will guide you to the right decision.

Summary

In this chapter we have discussed several aspects of manual construction that contribute to making the manual easy to use. Careful choice of format, including type size and style, headings, and the use of white space can make manual pages attractive and easy to read. Providing adequate referencing in the form of tables of contents, indexing, and cross-referencing (especially in troubleshooting sections) can make it easy for readers to find the information they need. Good design choices for the mechanics of the manual (page and cover stock, binding, and size), based on probable conditions of use, will help ensure that the manual you create will be read and referred to often. Care and attention given to these "surface" details will maximize the effect of the work you have put into writing, organizing, and creating good visuals.

5

Visuals

Overview

In an operator manual, as in a service manual, the visuals—photographs, drawings, charts, and tables—may be more important than the words. Clear, readable instructions and descriptions are necessary; clear visuals are vital. Most users, confronted with a manual for the first time, will leaf through the pictures long before they will read the words. When they read the words, they will rely on the pictures to help make unfamiliar terms clear: when told to "tap leg closures firmly until well seated," the new barbecue grill owner will look at the drawing to find (with relief) that the leg closures are merely the little plastic caps that go over the ends of the grill's tubular metal legs.

It is far easier to *show* than to *tell* how a machine works or how a maintenance procedure is performed. Of course, it is impossible for every owner to have an expert in residence to demonstrate procedures, so manufacturers provide operator manuals. Pictures provide the bridge between the words of the manual and the parts of the machine. Ideally, they help the reader locate a part, perform an adjustment—in short, get to know the machine. Too often, ill-designed, poorly labeled visuals frustrate the reader and present a poor image of the manufacturer. In this chapter we will discuss how to choose and design visuals and place them in a manual for maximum effect.

When to Use Visuals, and What Kind to Use

First of all, use as many pictures and drawings as you can: most manuals use far too few. Remember that the reader is not as familiar with the product as you are. Having written the manual, and (if you are an engineer-writer) possibly having designed the product, it is easy to forget the questions and confusions of the first-time user. Try to look at your product with fresh eyes—imagine it is your first look—and think about what drawings and photographs would help you to understand the instructions. This is simply another form of writing with the user in mind. In any case, don't skimp. Using fewer visuals may cut the cost of the manual, but it may also result in misused products, service calls, and less repeat business because users find the operator manual obscure.

Second, choose the best type of visual for your purpose and audience. You have three basic types to choose from: drawings, photos, and charts and tables. Charts and tables deal specifically with numerical information; we have discussed them in Chapter 3. Drawings and photos are both picture media, but they have important differences that affect the choice of which to use in a specific situation. Let's look at these in turn.

Drawings

Drawings are popular with manual writers for a number of reasons. You can show exactly what you want to show without having to deal with a clutter of extraneous parts. You can use a drawing to show normally hidden parts (as in a cutaway) or to show assembly sequences (as in an exploded view). (See Figures 5.1 and 5.2.)

Finally, drawings are often used because they are cheaper to reproduce than photos. A line drawing (or any purely black-and-white image—called *line art*) may be treated just like text by the printer; it can even be photocopied. A photograph, however, requires special treatment. The photo does not have just blacks and whites—it has continuous tones of gray (hence, photos are called *continuous-tone art*). (See Figure 5.3.)

Figure 5.1.   Cutaway View of a Valve. A cutaway view allows the reader to see the construction of an object by presenting it as if a portion were literally cut away, revealing the layers of composition. The cutaway view is particularly useful to show the interior of something that is not normally disassembled. *Source:* Drawing by Teresa Sprecher.

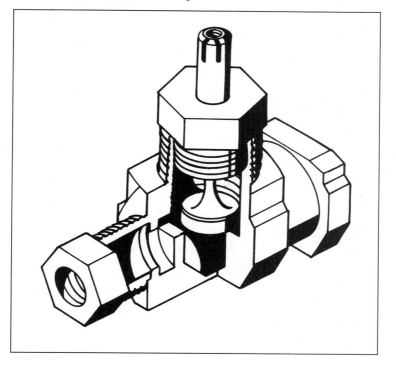

To reproduce such an image with black ink on white paper requires that the continuous-tone image be turned into blacks and whites. This is done by shooting the photo separately from the text through a *halftone screen*, reducing the photo to a pattern of dots of variable size and spacing. This dot pattern may then be treated as line art. (The dot effect can easily be seen by looking at a newspaper photograph under a magnifying glass.) (See Figure 5.4.)

The finer (and more expensive) the grid on the screen, the more detail is preserved. This extra step in the processing of photo-

Figure 5.2.   Exploded Diagram of a Valve. An exploded diagram allows the reader to see how the parts of an assembly fit together. It is most often used in conjunction with instructions for assembly or disassembly procedures. *Source:* Drawing by Teresa Sprecher.

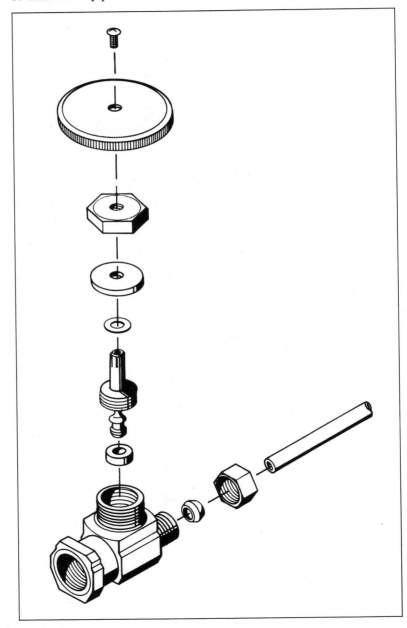

Figure 5.3.   Continuous-Tone Art. Note that continuous-tone art contains actual shades of gray, which cannot be reproduced by black ink on white paper. Special processing is required. *Source:* Courtesy of Engineering Publications Office, University of Wisconsin–Madison.

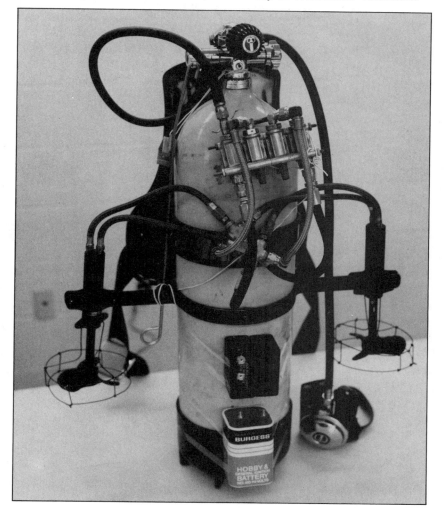

graphs costs money. However, this extra cost for a photo may be equalled by the extra expense of producing a drawing in the first place—especially if the drawing is designed especially for the operator manual and is not an engineering drawing that already exists.

Figure 5.4.   A Screened Photograph, or Halftone. The close-up shows the photo broken up into dots of various size and spacing. The normal view (Figure 5.3) illustrates how our eyes blend the dots to reproduce the original gray tones of the photograph. Because the screening process breaks the photo into a dot pattern, what was originally a continuous line (an object's edge, for example) becomes a discontinuous series of dots, and fine detail may be lost. The coarser the screen, the greater the loss of detail and clarity. *Source:* Courtesy of Engineering Publications Office, University of Wisconsin–Madison.

All of these advantages, however, stem from the fact that a drawing is *different* from an actual view of the product, and this fact can also present difficulties. A drawing requires interpretation; it is an abstraction. Many users find this translating hard to do, especially if they are not familiar with the product. The writer must always be aware of his or her audience and make choices accordingly. For example, it is usually a much better idea to provide a perspective drawing of a product to the new owner than to include an engineering diagram. (See Figures 5.5 and 5.6.)

Figure 5.5.   Perspective Line Drawing of a Valve. This drawing is an abstraction—it is not *really* the valve—but it is quite recognizable to an untrained eye. *Source:* Drawing by Teresa Sprecher.

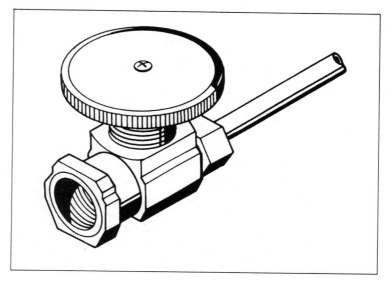

Unless your audience is technically sophisticated, leave the technical drawings on the drafting table. Even line drawings of perspective views may be difficult for some readers because of the lack of shadows, textures, and details. And the use of drawings to highlight a specific part without background clutter may become a disadvantage if the user cannot then recognize the part in context.

Photographs

The advantages of a photograph are virtually the opposite of those of a drawing. The chief advantage, of course, is realism—the photo looks like the product (if it is not a photo of a different model), and therefore recognition is easier. A photo shows parts in context, therefore making it easy to locate a particular item. And a photo does not require interpretation, as does a drawing; thus it is suitable for a nontechnical reader.

Figure 5.6.   Engineering Drawing of a Valve. This drawing is much more abstract and requires special training to interpret. *Source:* Drawing by Teresa Sprecher.

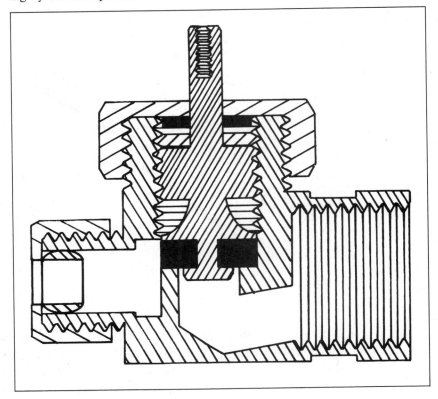

   Photos have two major disadvantages. First, because of their realism, they may be cluttered and not isolate the important part. Second, they cannot be used for views of hidden areas—you cannot very easily have a cutaway photo! (In Chapter 7, "Service Manuals," there is a cutaway view that is either a photo or a drawing made to look like a photo. It is, however, an unusual example.) The problem of eliminating clutter and isolating the important part may be dealt with in several ways. Two of the most common ways are to use contrasting arrows or outlining to point out the important parts and to airbrush out background distractions. (See Figures 5.7 and 5.8.)

Figure 5.7.   Example of the Relevant Area of a Photo Outlined in a Contrasting Tone. This technique is useful for combining the realism of a photo with the ability of a drawing to focus the reader's attention. *Source:* Reprinted by permission from *Operating Instructions*, IBM Correcting Selectric and IBM Selectric II, © 1973, International Business Machines Corporation.

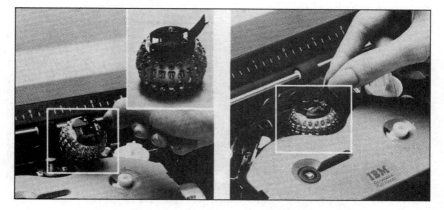

For other ideas, consult a commercial artist or photographer. The other disadvantage—realism—cannot be so easily overcome. If you need a cutaway view, use a drawing. However, when you do use photos, as far as possible show views that the user can actually see. In so doing, you will be making the most use of the photo's realism.

One other caution is in order here. For a photo to be of any use to the reader, it must be shot well and reproduced well. Nothing is more frustrating to a reader than to find that a figure is hopelessly out of focus or overexposed. If you have a photography department, or if you are both writer and photographer, the following suggestions may help to cut costs and make your photos more effective.

Photography Suggestions

1. Set up an area or a room in which you can control and vary the illumination with floodlights and light-reflecting boards. (Muddy and underexposed shots mean retakes.)

Figure 5.8. Example of the Relevant Area of a Photo Indicated by a Contrasting Arrow. Note how the arrow in the picture is aligned in the direction that the intended force is to be applied. *Source: Volkswagen Rabbit/Scirocco Service Manual, 1980 and 1981 Gasoline Models Including Pickup Truck,* © 1981, Robert Bentley, Inc. Used by permission.

2. Experiment with backgrounds and backdrops. For small products, use a table on which you can change the surface and the backdrops with changes of matte-surface posterboard or cloth. You will soon discover whether dark colors, shades of gray, or white work best to provide the contrast you are after. (A product painted blue will show up on a black-and-white photo as a different shade of gray than a product painted red or green.) Choose backgrounds and backdrops that do not "bleed" into the contours of your product. Make sure the background is light enough for the labels and numbers you want to add later.

    For large products, such as appliances or industrial or farm machines, outside photography may be necessary. If you have to shoot outside, keep background clutter to a minimum. Try to avoid glaring parking-lot surfaces, rough, stubbly, or dark corn fields, fences, utility poles, or building walls with strong design patterns. Avoid anything that draws attention away from the product or obscures its contours or key features. Or drop out the background entirely in the finished photo. If you can shoot inside and control illumination, a rotating platform that can be raised and lowered is a big help. It allows you to turn the product rather than having to rearrange lights and background as you shoot from different angles.

3. Polaroid photography is now sophisticated enough to provide ample clarity and definition for many photos of products. You can do your own photography, planning the shots to coordinate with the verbal text you have in mind. Some writers never write a line until the photos are done. The photos govern the text. Polaroid shots cost far less than studio shots, and exposure, focus, and camera angles can be checked on the spot.

4. If possible, work with an assistant. The assistant can turn the product around slowly until you get the angle you want and can switch lights, backgrounds, and backdrop materials. The assistant can be the "operator" of the product, that is, the hand that turns a knob or applies a tool, or the foot that steps on the brake.

5. Airbrush artwork is used to add highlights, reduce glare on metallic and shiny surfaces, and give depth to photos and drawings. Airbrushing photos, however, is expensive. Deere and Company has developed a cheap alternative to airbrushing that mimics its effects. Photographers spray white deodorant powder on areas that would ordinarily need airbrush work. The powder can then be wiped off with a soft cloth.

On the balance, our advice would be to use photographs whenever possible, unless for some reason a drawing would clearly be more suitable. People generally find photographs much easier and more helpful to use—provided, of course, that they are well done.

Designing Visuals

The first task in designing an effective visual, as in designing an effective piece of writing, is to figure out the purpose of the visual. What exactly do you want to show? Do you want an overall shot of the product simply to allow the reader to recognize the model you are going to discuss? Is the visual supposed to show the location of major subsystems or a single adjustment screw? Is the purpose to show how something is assembled or what it looks like in place? You must define as precisely as possible the job a particular illustration is to do. If you have trouble specifying one overriding purpose, maybe you are asking one picture to do the work of several. Consider using more than one visual if you have more than one purpose; otherwise, you may end up with an illustration that, in trying to do everything, does nothing effectively.

Once you have determined the purpose of the visual, you can begin to make the design choices that will ensure that the visual works—because you know what is important. The essence of good visual design may be summed up in the following rule: *Make the important things stand out.* For example, depending on what you consider to be important, you may decide to use a drawing rather than a photograph, or one view rather than another. If you are designing a table or a chart, you can tell how to set it up. As far as time and budget permit, make these decisions for each illustration individually. Don't use an old photo from another manual just because it is handy. In the long run, an illustration or table designed with a specific purpose in mind will better serve the user's needs and the company's interests.

The following general principles apply to any visual presentation—drawing, photo, or chart.

Make It Big

The user should easily be able to read and interpret the visual at normal reading distance. Remember that an operator manual may be used in less than ideal conditions. Depending on the product, the user may be reading the manual in a basement or a dimly lit barn. In any case, the user will probably be glancing from manual to product and back again. This makes it important for the user to be able to "find his or her place" in the visual rapidly—a maneuver that is much easier if the visual is large. In planning the layout, try to make spaces to fit the visuals rather than the other way around. Photo-reduction is a wonderful thing, but it can ruin a drawing or photo if used indiscriminately. In a drawing, for example, too much reduction can cause letters to fill in and closely spaced lines to run together. In a photo, depending on the coarseness of the halftone screen used to reproduce it, excess reduction can obliterate important detail.

Make It Simple

Without question, the most common fault of illustrations in operator manuals is that they are cluttered. A user will find it difficult to focus on (or even figure out) what is important in a visual overloaded with information. Illustrations and charts must be edited just like prose: figure out what the purpose of the visual is, then include only what is necessary to fulfill that purpose. You should not include everything you know in a visual, any more than you would in a paragraph. Simplify the visual presentation so that only essential items are included in detail, and nonessential items are either absent or merely suggested. (See Figure 5.9.)

If, despite your efforts to keep it simple, you still seem to have a complex illustration, consider splitting it in two—break the presentation down by systems, or show one overview and one or more close-ups.

This process of simplification is terribly difficult—there is a great temptation to include more than you should—but you will find that knowing the purpose of the visual will help enormously. If you have a clear idea of what you want the drawing, photo, or chart to accomplish, you can use that idea as a filter to screen out peripheral information.

Finally, use plenty of white space around and within the visual. This alone will help to reduce visual clutter and make the visual more inviting to use.

Figure 5.9.   Example of a Photo with the Background De-emphasized. Note how fading out the rest of the engine permits the reader to see the placement of the water pump in relation to the rest of the engine while still focusing on the pump itself. *Source: Volkswagen Rabbit/Scirocco Service Manual, 1980 and 1981 Gasoline Models Including Pickup Truck,* © 1981, Robert Bentley, Inc. Used by permission.

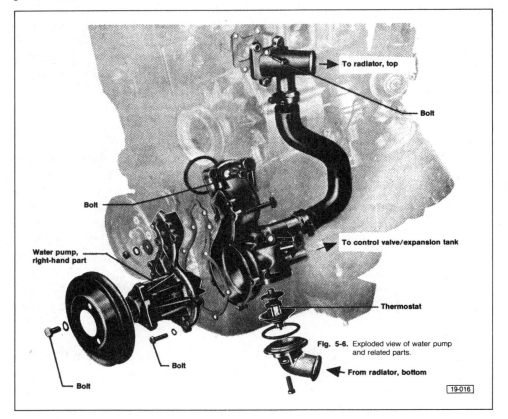

Fig. 5-6. Exploded view of water pump and related parts.

Make It Clear

Be sure that each drawing, photo, table, and chart has a title that tells what it shows as well as a figure number or table number. For example, "Figure 3, Location of Idle Adjustment Screw" is much better than just "Figure 3." Label the parts of the illustration adequately—and as close as possible to the parts referred to. The object is to cause the reader's eye to jump back and forth as little as possible. For this reason, it is better to put the labels in words on the drawing or photo itself rather than use callouts (numbers or letters listed in a key elsewhere). Sometimes direct labeling is not possible or desirable, as, for example, if labels would clutter the illustration, or if the manual is to be prepared in more than one language.

Be sure that the visual message supports the verbal message of the illustration. In other words, make sure that the important part really does stand out. Use white space and the direction of lines in the drawing or photo to emphasize the important parts. Do not rely on labels or explanations to overcome the effects of poor visual design: people react more strongly to the visual image than to the words, just as they react more strongly to body language and tone of voice than to the substance of a conversation.

Integrating Visuals with Text

Always plan your visuals at the same time as you plan your text. While you work on the outline of your text, be thinking about where to include visuals. Make sketches as you develop a rough draft. This will ensure that text and illustrations will balance and support each other. This principle is carried out in the ''illustruc-tion'' method developed by Deere and Company in which each block of text is written in terms of an illustration. (See Figure 5.10.)

As in the prose of the text, try to follow a general-to-specific order of illustrations: start with an overview, then move to subsystems and close-ups of individual parts.

Always refer to your visuals in the text, and always place the visual as soon as possible after the first reference to it.

Figure 5.10.   Sample Page from a Deere and Company Manual Showing the "Illustruction Method." Note how each photo is keyed to a corresponding block of text, which is kept quite short. The small numbers in the lower right-hand corners of the blocks are the computer "addresses" of the blocks. Thus, relevant segments may easily be used in different manuals. Note also that the photos are on the outside of the page, making it easy to look back and forth from book to product. *Source:* John Deere *Operator Manual* (OM RW 15455) Issue A1, Waterloo Tractor Works. Used by permission.

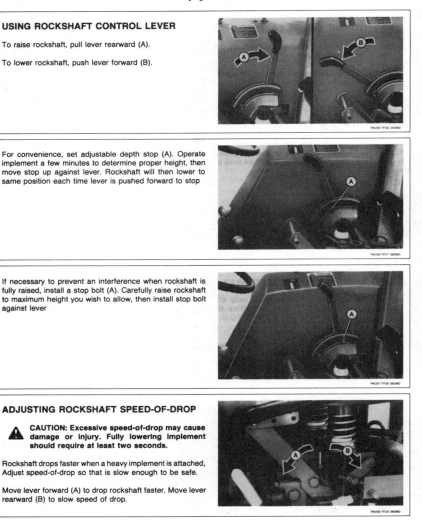

Review Checklist and Exercise

Checklist

The following is a list of questions to ask yourself about the visuals you have included in your manual.

- ☐ Have I planned what visuals to include from the user's point of view? Have I tried to look at my product with a fresh eye?
- ☐ Have I defined carefully the purpose of each visual?
- ☐ Have I chosen the type of visual to use (drawing or photo) based on that purpose?
- ☐ Have I designed my visuals to be big, simple, and clear?
- ☐ Have I taken complex visuals and tried to break them down into simpler illustrations?
- ☐ Have I used adequate white space?
- ☐ Have I carefully labeled the parts of the visuals?
- ☐ Have I tied the visuals clearly to my text?

Exercise

The sample instructions that follow might appear in an operator manual for a sewing machine. For those of you not familiar with sewing machines, here is a brief description. The balance wheel is a large flywheel mounted on the right side of the machine. As the needle (left) goes up and down, the balance wheel turns. A spool of thread is mounted on top of the machine, and the thread goes through various devices to maintain proper tension and finally goes through the needle itself. Underneath the base of the machine is a small metal spool of thread called the bobbin, which rests in a removable metal case. Thread from this bobbin is fed up through a hole in the bed of the machine and interacts with the needle-controlled thread to produce stitching.

Example 5.1 is an explanation of how to remove the bobbin. Read it and do the following:

Example 5.1. Sample Instructions

> To remove the bobbin:
>
> Turn the balance wheel until the thread take-up arm is at the top of its travel. The needle may not be in the highest position. Lift the hinged portion of the bed, and with the left hand reach under the machine and open the latch of the bobbin case. Still grasping the latch, pull the case away from the machine to your left. Remove case (with bobbin inside). To release the bobbin from the case, hold the bobbin case in your right hand and, with your left hand, close the latch. Turn the open end of the bobbin case downward, and the bobbin will fall out.

1. Write down the questions that occur to you as you read.
2. Check the ones you think could be answered by a visual.
3. List and describe the visuals you would include. Be sure to address these questions for each one:
   a. What will it show?
   b. Will it be a drawing (what sort?) or photo? Why?
   c. What will you label, and how?
   d. Where in relation to the text will it appear?

Summary

As you can see, choosing and designing good visuals for an operator manual follows much the same process as the writing itself: you must determine your purpose and audience, then choose the visuals in terms of those. You must select and organize information that is presented visually, as you must for information that is presented verbally. You must develop sketches or visual drafts, and, finally, you must "edit" your tables and illustrations according to the rules for visual clarity. The process is the same because the function is the same— words and pictures combine to help the reader learn how the product works and how to operate and maintain it properly; only the language is different.

6

Safety Warnings

Overview

Increasingly, manufacturers have become concerned with product liability. Therefore, they have been paying more attention to the role of instruction manuals and safety warnings in ensuring safe use of products, especially because these documents may be used as evidence in litigation. In this chapter we first present a short background in product liability law and discuss the duty to warn. Next, we offer guidelines for designing effective safety warnings, both for labels on the product itself and for use in the manual. (The task of developing label designs often falls to the manual writer; in any case, all safety messages, whether on labels or in the manual, must be coordinated.) We discuss the pros and cons of standardized warnings and safety labels. Finally, we present guidelines for related material included in the manual, such as disclaimers and documentation that the user has received the manual.

Product Liability Background

The rise in the consumer movement and the increase in product liability suits in recent years have placed a greater responsibility on manufacturers to make their products safe and to provide both proper instructions for their use and proper warnings of hazards. The area of product liability is extremely complex, and the material in this chapter is necessarily general. As you write a manual, you should work closely with your company's product safety division and legal counsel to ensure that your part in the manufacturing process meets legal requirements. (See also Chapter 1, "Planning.")

Product liability litigation may be based on one of three legal concepts: *negligence, breach of warranty*, and *strict liability in tort. Negligence* means that the manufacturer did not exercise reasonable care in the manufacture or marketing of a product, resulting in an unreasonable risk to the user. *Breach of warranty* means that the product did not do what the manufacturer said it would—but note that the warranty can be either express or implied. An implied warranty may be present in instructions for the product's use, even if the matter in question is not included in the written (or express) warranty. *Strict liability in tort* means that, if a product is defective in manufacture or marketing, the manufacturer may be liable for damages even if the manufacturer was *not* negligent, was unaware of the defect, and made no claims about the product's performance. Strict liability in tort concentrates on the product, not on the care with which the manufacturer operated. Whatever legal concept is used, four things must be true for a product liability suit to be successful.[1]

- The product must have a defect.
- The defect must be present when the product leaves the control of the manufacturer.
- Injury or damage must be incurred.
- The injury or damage must have been caused by the defect.

A product defect may be a design defect, a manufacturing defect, a packaging defect, a marketing defect, etc. In this chapter we will deal only with defects involving instructions and warnings. These fall into three categories:

- Failure to warn at all of risks or hazards present in the product
- Failure to warn adequately of such risks or hazards
- Failure to provide appropriate and adequate instructions for use of the product

Let's look more closely at this duty to warn.

Who Must Be Warned

The law states that the manufacturer has a duty to warn anyone who might reasonably come into contact with the product. Thus, the manufacturer may be held liable for injury to someone other than the person who actually bought the product—for example, an employee of a company that bought an industrial cleaning product or the child of a consumer who bought an electric coffee grinder. Especially if the likelihood of injury is great or the potential injury is serious, the manufacturer may be required to include a warning directly on the product itself. The situation for the technical writer is further complicated by the fact that the potential users of a product may comprise a very diverse group. (See Chapter 2, "Analyzing the Manual User.") You must consider the possible range in terms of age, sex, expertise, familiarity with product, even literacy. In *Hubbard-Hall Chemical Co.* v. *Silverman*, the court held that a written warning was not adequate because it failed to provide for illiterate users.[2]

What Must Be Warned About

The manufacturer has a duty to warn potential users of dangers present in the nature of the product in normal use *and* in foreseeable misuse of the product. Thus, a manufacturer of chlorine-containing laundry bleach has a duty to warn of potential skin irritation, since this is a risk inherent in using the product to bleach clothes. However, the manufacturer may also have a duty to warn against mixing the product with ammonia (as someone might do if using the bleach as a household cleaner), which produces deadly chlorine gas, because this is a foreseeable misuse.

The manufacturer has no duty to warn of open and obvious dangers—that a knife cuts, for example. However, you must be sure that the danger is obvious to the user. A danger that is open and obvious to you, who are thoroughly familiar with your company's products, might be unknown to the user. For example, many people are unaware that burning charcoal emits carbon monoxide gas, which can be deadly with inadequate ventilation. Anyone concerned with the manufacture of charcoal briquettes surely knows this, yet every year the news includes reports of people dying while using a charcoal grill in a trailer or closed garage.

Even if the danger is open and obvious, the manufacturer may have a duty to warn if the user may not be aware of the extent or degree of danger. For example, a person using tile adhesive labeled "flammable" probably would not smoke while using the product, but he or she might well not realize the danger posed by pilot lights, especially in remote areas of the home.

How Long the Responsibility to Warn Continues

The manufacturer has a continuing duty to warn of hazards connected with the product, even if the hazard is discovered after the sale of the product. (Note, however, that, in general, litigation involving the continuing duty to warn has been based on negligence rather than on strict liability in tort.[3]) The manufacturer may, in extreme cases, need to recall the product; in other cases, it may be sufficient to provide replacement parts or a warning of the hazard with instructions on how to avoid it. In any case, you should have a system for notifying owners of changes in recommended use or service procedures or of modifications that should be made to the product.

Older products are of particular concern to manufacturers, because safe use of older products often depends heavily on the operator's being aware of hazards and taking steps to avoid them. Manufacturers have no guarantee that anyone but the original owner of a product will receive the instruction manual. In addition, safety equipment on older products was often easily removed or bypassed. For example, a recently manufactured drill press might be designed so that the power cannot be turned on unless a movable safety shield is in place. An older drill press might include the safety shield, but make its use optional. Courts have awarded damages to persons injured by machinery from which safety shields had been removed because removal of safety equipment was considered to be a foreseeable misuse.[4] These cases point to the need for warnings placed on the product itself, as well as in the instruction manual—particularly if your company's product is likely to be resold or used by persons other than the original purchaser.

What Is an Adequate Warning?

To be considered adequate, a hazard warning must do four things:

- Identify the gravity of the risk
- Describe the nature of the risk
- Tell the user how to avoid the risk
- Be clearly communicated to the person exposed to the risk

All four of these elements must be present. For example, in a case involving asbestos fiber hazards, the court found a warning inadequate because it did not state the severity of the risk and the nature of the hazard.[5] The warning read as follows:

> This product contains asbestos fiber. Inhalation of asbestos in excessive quantities over long periods of time may be harmful. If dust is created when this product is handled, avoid breathing the dust. If adequate ventilation control is not possible, wear respirators approved by the U.S. Bureau of Mines for pneumoconiosis-producing dusts.

Actually, this particular warning could serve as a model of how *not* to write a warning. We will return to it later as an annotated example.

Strategies for Warning and Instructing

Warning Labels on the Product

When you are developing warning labels to be placed on the product itself rather than in the manual, you should try to design labels that follow these guidelines:

- Make your warnings consistent.
- Never mix general instructions with warnings.
- Follow existing guidelines.
- Make sure warnings meet all applicable standards.
- Place the warning near the hazard.

- Make sure the warning is readable.
- Make sure the label is durable.

Let's look at each of these guidelines in turn.

Make Your Warnings Consistent   This principle applies not only to warnings, but, as we have seen, to general instructions as well. (See Chapter 3, "Organization and Writing Strategies.") By following a consistent format, you set up a pattern of expectations in your readers—they expect to find the same sort of information in the same places from warning to warning. As long as you remain consistent, you remove from your readers the burden of figuring out the *type* of message you are giving them, allowing them to concentrate their attention on the message itself.

Never Mix General Instructions with a Warning   A warning may be overlooked or the seriousness of its message diluted if it is combined with general instructions. This is generally more of a problem in instruction manuals, but sometimes labels contain a mixture of warning information and other information that would normally also be included on a product label. For example, a label might read "WARNING: steam under pressure can cause severe burns. Release pressure before removing radiator cap. Keep radiator filled to appropriate level with 50/50 mixture of recommended coolant and water." The last sentence of that warning is related to the hazard, since an underfilled radiator will build up more pressure than a filled one, but it does not relate to the immediate hazard of escaping steam. The shorter the better—as long as all necessary information is given about the nature and severity of the risk and the means to avoid it.

Follow Existing Guidelines   Two particularly helpful sets of guidelines are put out by the FMC Corporation and by Westinghouse. These may be obtained at a reasonable charge from those companies.[6] The guidelines suggest designing warning labels consisting of three parts:

- A signal word and color to convey the severity of the hazard

- A symbol or pictogram showing the nature and consequences of the hazard
- Words to describe how to avoid the hazard

Figure 6.1 shows some sample warning labels.

These guidelines suggest that a signal word be used consistently to convey a particular level of risk:

- DANGER (red): the hazard or unsafe practice *will* result in severe injury or death.
- WARNING (orange): the hazard or unsafe practice *could* result in severe injury or death.
- CAUTION (yellow): the hazard or unsafe practice could result in *minor* injury or property damage.

Other instructions or cautionary statements pertaining to product damage (e.g., "make sure there is enough oil in your lawn mower's crankcase") should use other signal words, such as *notice*.

The symbol or pictogram should depict the nature of the hazard and its consequences, preferably by showing both the machine part and the body part. FMC has developed a series of pictograms to cover various common hazards. (See Figure 6.2.) The use of symbols or pictograms in addition to a verbal message is particularly important if the product in question may be used by persons who are illiterate (including children) or who do not speak the language in which the warning is written.

The final element of the warning label, the verbal message telling how to avoid the hazard, must be clearly communicated to the reader. The wording should be simple, direct, and active. Do not be afraid to word a warning strongly and specifically. Do *not* say: "may result in bodily harm"; *do* say: "can amputate fingers."

At this point, let's return to the warning about asbestos fibers and analyze it in terms of these verbal guidelines.

This product contains asbestos fiber. Inhalation of asbestos in excessive quantities (*How much is that?*) over long periods of time (*How long?*) may (*Possible, not probable?*) be harmful (*How harmful?*). If dust is created when this product is

Figure 6.1. Sample Warning Labels. *Source: Product Safety Label Handbook* © 1981, Westinghouse Electric Corporation. Used by permission.

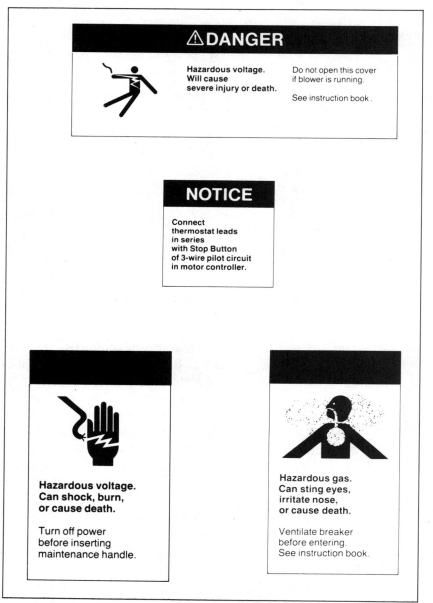

Figure 6.2.   Sample Pictograms. *Source: Product Safety Sign and Label System*, 3rd Edition, 1980, FMC Corporation. Used by permission.

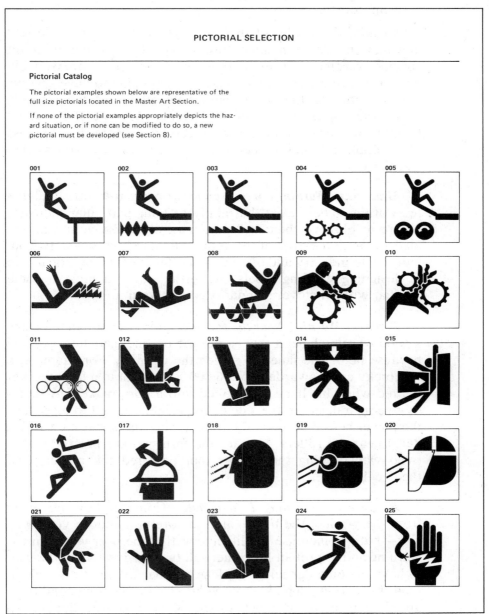

PICTORIAL SELECTION

Pictorial Catalog

The pictorial examples shown below are representative of the full size pictorials located in the Master Art Section.

If none of the pictorial examples appropriately depicts the hazard situation, or if none can be modified to do so, a new pictorial must be developed (see Section 8).

handled, avoid breathing the dust (*How?*). If adequate ventilation control is not possible (*What's adequate?*), wear respirators . . .

Clearly, the reader of this message is given very little real information. Some manufacturers fear that writing forceful, specific warnings will make their products seem unreasonably dangerous. In general, potential buyers are glad to have specific information, and the courts have held numerous times that a vague warning is an inadequate warning. A successful product liability suit can cost a company a good deal more than the loss of a couple of sales to persons "scared off" by well-written warnings.

Make Sure Warnings Meet Applicable Standards   If standards for warning labels exist that apply to your particular industry, failure to observe them will automatically make your warning inadequate in the eyes of the court. Standards vary widely from industry to industry and often do not offer specific, practical help in designing warning labels. Nevertheless, you must become familiar with whatever standards govern your area.

Place the Warning Near the Hazard   It is not enough to design a good warning label if the user of the product gets hurt anyway because he or she failed to notice the warning. For example, a Kentucky court awarded a $266,000 settlement to a farm worker injured when his hand was pulled into a corn picker.[7] A three-by five-inch decal on the side of the picker warned to disengage the power before cleaning the rollers—but there was no warning decal near the rollers. Neither the farm worker nor the owner (who was running the tractor that powered the picker) saw the decal.

Be Sure the Warning Is Readable   Consider how far away from the product the user will be before you choose your type size—both to increase the likelihood that the user will read the warning and to protect his or her safety. You don't want the user to have to come dangerously close to a hazard just to be able to read the warning! Think about what kinds of lighting conditions are likely when the product is used: will the light be dim (as in a poorly lit basement or barn) or glaringly bright (as in a sunny corn field)? Choose colors, and label material accordingly. What is the viewing angle? Will the user be able to look directly at the warning where you plan to place it? If the angle is too severe, the words may not be readable. Try to foresee all the likely conditions of use for your product—and *then* make your design decisions.

Make Sure the Label Is Durable   A carefully designed warning label is useless if it dissolves in the rain, is rubbed off in a few weeks of use, or is totally obliterated by a few smudges of oil or dirt. In choosing the materials you use for your labels, you must consider the material to which the label will be attached, the length of time it will have to last, and the kind of treatment it will receive. Be sure that the materials match—i.e., that the base material and the adhesive are compatible and of comparable durability (there is no point, for example, to a long-lived base material and a short-lived adhesive). Be sure that the ink used to print the message will set well with the base material. Make these decisions early, and consult with suppliers for specifications on various materials.

Instructions and Warnings in Manuals

Instructions for proper use of a product must be clear, readable, and understandable by readers. Follow the guidelines provided in the rest of this book for creating good instructions. Clear and complete writing is the best way to avoid inadequate instructions.

Warnings included in the text of an instruction manual must follow many of the same guidelines as labels on the product itself. They must be

- Internally consistent
- Consistent with the labels on the product itself
- Consistent with applicable standards

You may have many more warnings in the text than labels on the product, but be sure that every DANGER-level warning in the text (hazard or unsafe practice that will result in severe injury or death) has a corresponding label on the product. Be careful that you do not overuse the DANGER designation—too many

Figure 6.3. **Example of Safety Warnings Embedded in Text.** *Source:* Unidentified. Safety warnings are insufficiently highlighted and mixed in with general instructions. Our annotated version (facing page) shows that almost half the text contains warnings to protect both operator and onlookers.

BEFORE WELDING –

The powerline welder takes the power from the supply line and converts it to a lower voltage, higher current output suitable for welding. The ground clamp and electrode holder form the ends of the welding circuit. When an electrode is inserted in the holder and struck on the work, the arc completes the circuit.

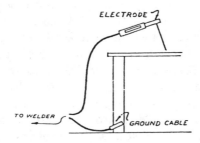

ELECTRODE

TO WELDER

GROUND CABLE

Wear a proper welding helmet. Never, under any circumstances, look at the arc without proper protection. Warn anyone nearby and keep pets away. Although not likely to cause permanent injury, even a few arc "flashes" may be the cause of headaches and pain in the eyes. A good deal of smoke, hot sparks and spatters of hot metal are given off when welding. It is only common sense to weld in a well ventilated area, away from combustibles, and wearing protective clothing. Pocket flaps and pant cuffs should be turned down. Shirt sleeves and collars should be buttoned to cover the exposed skin. Besides protecting from hot particles, this will prevent arc burns to the skin. These are like sun burns and can be especially severe when using the carbon arc torch.

When care is taken to see that the work area is clean and safe, the welding operator can concentrate his attention on the job.

WELDING

Get a piece of scrap steel 1/4" or 3/8" thick to practice on. The proper heat settings can be taken from the name plate instructions. On some models, the rod sizes are also given on the name plates, together with a gauge for metal thickness.

Insert the plugs in the machine, using a slight twisting motion to ensure good contact.

NEVER remove or change the plugs while the operator is welding because they will burn or pit.

Fasten the ground clamp to the steel and put a 1/8" 6013 electrode in the electrode holder. Set the heat at about 130 amps. Turn the welder switch on and scratch the rod lightly across the steel (helmet down) just as you would strike a match. Draw back slightly on the rod to lengthen the arc. Try to maintain the arc by keeping the rod a short distance from the base metal. Feed the rod downward as it melts away, otherwise, the arc will lengthen and go out. If the rod is held too close to the work, it may cause the arc to go out and the rod will "stick". If the rod should stick, it may be twisted free.

of them may dilute the impact. If you find yourself writing an inordinate number of DANGER warnings, it may indicate that the product is unreasonably dangerous and should be redesigned.

Figure 6.3. *(continued)*

BEFORE WELDING –

The powerline welder takes the power from the supply line and converts it to a lower voltage, higher current output suitable for welding. The ground clamp and electrode holder form the ends of the welding circuit. When an electrode is inserted in the holder and struck on the work, the arc completes the circuit.

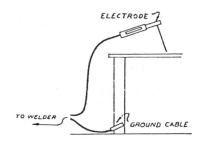

ELECTRODE

TO WELDER

GROUND CABLE

EMBEDDED SAFETY MESSAGE

Wear a proper welding helmet. Never, under any circumstances, look at the arc without proper protection. Warn anyone nearby and keep pets away. Although not likely to cause permanent injury, even a few arc "flashes" may be the cause of headaches and pain in the eyes. A good deal of smoke, hot sparks and spatters of hot metal are given off when welding. It is only common sense to weld in a well ventilated area, away from combustibles, and wearing protective clothing. Pocket flaps and pant cuffs should be turned down. Shirt sleeves and collars should be buttoned to cover the exposed skin. Besides protect-ing from hot particles, this will prevent arc burns to the skin. These are like sun burns and can be especially severe when using the carbon arc torch.

When care is taken to see that the work area is clean and safe, the welding operator can concentrate his attention on the job.

WELDING

Get a piece of scrap steel 1/4" or 3/8" thick to practice on. The proper heat sett-ings can be taken from the name plate inst-ructions. On some models, the rod sizes are also given on the name plates, together with a gauge for metal thickness.

Insert the plugs in the machine, using a slight twisting motion to ensure good contact.

NEVER remove or change the plugs while the operator is welding because they will burn or pit.

EMBEDDED SAFETY MESSAGE

Fasten the ground clamp to the steel and put a 1/8" 6013 electrode in the electrode hold-er. Set the heat at about 130 amps. Turn the welder switch on and scratch the rod lightly across the steel (helmet down) just as you would strike a match. Draw back slightly on the rod to lengthen the arc. Try to maintain the arc by keeping the rod a short distance from the base metal. Feed the rod downward as it melts away, otherwise, the arc will lengthen and go out. If the rod is held too close to the work, it may cause the arc to go out and the rod will "stick". If the rod should stick, it may be twisted free.

Never mix safety warnings with ordinary instructions, and never bury warnings in the text in such a way that they might be missed. (See Figure 6.3.) Make sure warnings included in the text stand out and are easily readable.

Standardized Warnings and Labels

"Why not make all labeling and safety instructions uniform? Then there would be no mistakes." Many products resist neat categorizing or are composed of components not all of which are suitable for standardized warnings and labels. You have only to read the safety instructions for three or four different brands of baby toys, riding mowers, or power tools to see how different manufacturers vary in their choices for warnings and labels. Even so, some standards exist, mostly for narrowly defined and very specific products, such as hazardous industrial chemicals.

Benefits of Standardized Warnings

Standardized warnings have a number of advantages:

- Warnings and word choice have been subjected to committee review.
- Warnings represent a composite of the judgments of safety specialists in specific industries.
- Warnings are backed up by combined experience of many companies who make the product.
- The accumulation of case law and liability experience helps to identify trouble areas.

Those who favor this committee approach to creating standardized warnings also believe that repetition of the same warning symbol or words is itself a plus. They argue that confrontation with the same warning again and again serves to reinforce the message and to promote instantaneous recognition.

Limitations of Standardized Warnings

Checklists or "cookbook" warnings also have drawbacks, largely because the cookbook warning is only as good as the cooks who produced it. Even committee-designed warnings can reflect blind spots or errors in judgment, and a bad standardized warning can gain widespread use simply because it is standardized. Further, uncritical reliance on predesigned warnings can be deceptively easy and sometimes even dangerous. If you begin to ''plug in'' standard warnings without paying attention to the special characteristics of your product, safety problems may begin to slip by. Your product may only seem to be the same as others, when in reality you need to think creatively about its hazards.

Esthetic Trade-Offs

Some industries resist the color, size, or wording of standardized labels on esthetic grounds. For example, kitchen ware and appliances whose color and design are part of the sales appeal may suffer esthetically from the warning label. Manufacturers may choose to emboss glass or use an extruded plastic label rather than stick a blazing red label on an avocado green product. Trade-offs between esthetics and standardized warnings are common where sales of a product depend heavily on eye appeal.

Safety Cartoons

A difficult issue is the use of safety cartoons. The main attraction is that they are eye-catching. Most people will look at a cartoon in preference to reading a written warning. However, in our view, cartoons have a number of serious drawbacks, and we do not recommend their use. The shortcomings of safety cartoons include the following:

- They may dilute the warnings in the text—especially if the warnings incorporate pictograms, as we recommend. You cannot include a cartoon for every warning, but including cartoons for only some may imply that the other warnings are trivial.

- They may seem to treat a heavy subject lightly and thus undermine their own purpose. The cartoon format may remind product users of Saturday morning TV cartoons, in which terrible things happen to the cartoon characters without rendering them any permanent harm.

- They are very hard to design well. To effectively focus attention on a hazard, a cartoon must be extremely simple and uncluttered and must make the hazard itself clear and obvious.

- They may imply that you are talking down to readers, thus alienating them and making it less likely that they will read (and heed) other warnings or instructions for safe practice.

Given all these shortcomings, and given the availability of effective symbols and pictograms (which are also eye-catching), we recommend that safety cartoons be avoided.

Other Safeguards for the Manufacturer

Two other areas exist in which the manufacturer may further protect against product liability suits: front matter in the manual and documentation that the user has received the manual. Both these areas are legally important in proving that the user of the product got the proper instructions for safe use of the product.

Front Matter

Every instruction manual should be dated and should tell exactly what model product it covers and what previous books, if any, it replaces. The manufacturer is therefore protected from the claim that the user did not know that it was the wrong book for the product. Second, the front matter should include appropriate disclaimers. These may or may not protect the manufacturer in a given situation, but they cannot hurt. The disclaimers should state that:

- No warranties are contained in the manual other than . . . (state whatever warranty is appropriate for the product)
- The instruction book does not alter any agreement for division of responsibilities worked out between the manufacturer and the dealer—so that the manufacturer is not held liable for something the dealer should have done.
- The information in the manual is not all-inclusive and cannot cover all unique situations.

Documentation

The manual should include, in a conspicuous place (such as inside the front cover), some documentation that the buyer received the appropriate instruction manual with the purchase of the product. This may take the form of a postcard filled out and signed at time of sale and sent to the manufacturer, or some other form, but it should be included—especially in any manual for a product that is inherently hazardous. The manufacturer then has proof that it fulfilled its duty to provide instructions and warnings at least to the immediate purchaser. This does not, of course, guarantee that the information will be passed along to other users.

Review Checklist

Here is a list of questions to ask yourself about the safety warnings and messages that will go with your product. If you can answer yes to all the questions, you have probably done an effective job of helping to protect your product's users and your company.

- ☐ Have I identified and warned about all the hazards connected with my product?
- ☐ Have I looked at my product through the eyes of a first-time user?
- ☐ Have I anticipated foreseeable misuses of my product?
- ☐ Have I included warning labels on the product itself for severe hazards and for hazards connected with a product likely to be resold or used by someone without access to the instruction manual?
- ☐ Do my warnings include all four elements of an adequate warning?
- ☐ Do my warning labels follow a consistent format?
- ☐ Have I separated warnings from general instructions?
- ☐ Have I followed existing guidelines?

☐ Do my warnings meet applicable standards?

☐ Have I placed my warning labels near the hazards they warn against?

☐ Will the warning labels be readable during normal use of the product?

☐ Are the labels durable? Will they last as long as the product does?

☐ Have I included appropriate warnings in the text of the instruction manual?

☐ Are these consistent with any labels on the product itself?

☐ Do they stand out from the rest of the instructions?

☐ Have I included appropriate information in the manual about the model(s) it covers, the book(s) it replaces, or the date it becomes effective?

☐ Have I included appropriate disclaimers?

☐ Is there a means to document that the buyer received the manual?

Summary

The rapid increase in product liability litigation in the last decade and the case-law approach to defining product defectiveness have without question put a tremendous responsibility on manufacturers to evaluate product safety efforts. This situation has led to the development of much more sophisticated and standardized practices in writing instructions and safety warnings and thus raised the quality of these materials in general. The role of technical writers has become more important and more respected—and the job has become harder. The guidelines presented in this chapter should help you to design effective warnings and write helpful instructions to ensure that people use your company's product safely.

References

1. Ross, Kenneth. "Legal and Practical Considerations for the Creation of Warning Labels and Instruction Books." *Journal of Products Liability* 4(1981), pp. 29–45. See also "Section 402A," *Restatement of the Law Second, Torts 2d*. Philadelphia: American Law Institute Publishers, 1965.

2. *Hubbard-Hall Chemical Co.* v. *Silverman*, 340 F 2d 402 (1st Cir. 1965).

3. Ross, "Legal and Practical Considerations," p. 37.

4. *Craven* v. *Niagara Machine & Tool Works, Inc.*, 417 N.E. 2d 1165 (Ind. App. 1981).

5. *Borel* v. *Fibreboard Paper Products Corp.*, 493 F 2d 1076 (5th Cir. 1973); Cert. denied, 419 US 869 (1974). See also Ross, "Legal and Practical Considerations," p. 35.

6. *Product Safety Sign and Label System*, 3rd ed. Santa Clara, CA: FMC Corporation, Central Engineering Laboratories, 1980. *Product Safety Label Handbook*. Trafford, PA: Westinghouse Electric Corporation, Customer Service Section, Westinghouse Printing Division, 1981.

7. *Moore* v. *New Idea Farm Equipment Co.* 78-C1-069 (Lee County, Ky. Cir. 1981).

7

Service Manuals

Overview

Although service manuals employ many of the same principles and techniques as operator manuals, the two forms differ. A service manual has a much more specialized audience and purpose, and this difference is reflected in text and design. This chapter will discuss these differences, as they are reflected in content, style, visuals, and mechanics. Certainly, the principles of good verbal and visual design outlined elsewhere in this book will still apply: instructions should be presented in parallel form, visuals should be clear and easy to read, and so on. But the application of these principles will differ, and this chapter will present detailed guidelines for putting the principles to work in the specialized context of the service manual.

How Service Manuals Differ from Operator Manuals

Purpose

The purpose of a service manual is very different from that of an operator manual. An operator manual is intended primarily to give clear instructions for a product's use and care. It introduces a new user to a product and explains what the product is for and how to make it work. An operator manual may give simple maintenance procedures, such as how to clean the cabinet of a TV set or how to change the oil in a lawn mower, but such instructions usually cover only the most basic operations. For anything complicated, the user is usually referred to "an authorized service representative." A service manual, in contrast, is what the authorized service representative uses. The purpose of a service manual is to explain in detail the repair and maintenance of a product—to explain, for example, how to clean and adjust a faulty tuner in a TV or how to overhaul the engine of a lawn mower. Normally, a service manual will assume that the reader is familiar with the product, knows how to operate it properly, and needs specialized information only.

In addition, the service manual often serves as the "textbook" for training technicians. In factory training programs and in technical schools, students learn by doing. A student in a transmissions class at a technical college, for example, will learn how to repair auto transmissions by working on one chosen from a particular make and model of car. The service manual for the car will be the primary resource for the student learning how to do a particular procedure.

Audience

The audience for a service manual is also very different from that for an operator manual. As we have already discussed, the audience for an operator manual may be anyone from a professional user who is technically sophisticated to a member of the general public who perhaps has never even seen the product before, much less used it. The audience for a service manual, in contrast, is almost always technically sophisticated and very familiar with the product. This audience will most often consist of professional repair and service technicians but may also include knowledgeable amateurs—the do-it-yourself auto mechanic, for example.

Content

What a Service Manual Does Not Contain

Unlike an operator manual, a service manual will usually not include any introduction to the product. It will not cover what the product looks like, what its major components are, what it is used for, or what its capabilities are. Since the reader is assumed to be familiar with the product, this kind of information is not considered necessary. If the reader of the service manual is of the "knowledgeable amateur" category, he or she may need to rely somewhat on the operator manual for this kind of background information.

What a Service Manual Does Contain

Naturally, the precise content of the service manual will vary with the product, but certain categories of information will appear in all service manuals. These include the following:

- Specifications for the product, including capacities for lubricants, cooling agents, etc., and recommendations for lubricants and cleaning agents to be used
- Technical background on the function and operation of the product or its systems
- Routine maintenance procedures and recommendations for service intervals
- A troubleshooting guide
- Repair procedures
- Model change information
- A parts catalog (this may be a separate publication)

This chapter will discuss each of these categories in detail. First, however, we would like to make some general remarks about how these different kinds of information are treated in a service manual. The information in a service manual will naturally be far more complete than in an operator manual. More complete explanations will be given, and more complex procedures will be described. These procedures may require specialized tools. It is common practice for factory-produced service manuals to refer to tools by factory numbers instead of by generic names (e.g., "VW 558" instead of "flywheel holding fixture"). The writer of a service manual must be aware that *not all* users of the service manual will be authorized, factory-trained technicians. It might appear that using a specialized terminology for specialized tools would discourage amateurs; however, it is more likely that the amateur will simply find some other way to do the procedure (e.g., using a screwdriver rather than a snap-ring pliers). Example 7.1 shows how this issue has been addressed in one auto service manual. If a specialized tool is *necessary* for a given procedure for safety reasons or to avoid damage to the product, note this fact and refer to the tool by name as well as number.

Safety information in general is important in a service manual. The more complex procedures covered in the service manual are often also more inherently hazardous. In addition, the technician may grow careless through familiarity with the hazards and not take proper precautions. Or the technician, having performed similar procedures many times, may not read instructions closely and may miss vital safety information unless it is prominent. All of the guidelines given in Chapter 6 on safety warnings also apply to service manuals.

Example 7.1.   Accommodating Nonprofessional Users

3. Install the flywheel holding fixture on the pressure plate assembly, as shown in Figure 10-3. Alternatively, you can use coat hanger wire to bind a bolt hole in the pressure plate assembly to a bolt hole in the engine block's transaxle mounting flange.[1]

Specifications

Service manuals will contain detailed information on specifications and tolerances. Operator manuals will contain some of this same information, but only what is necessary for routine maintenance by the owner. For example, the operator manual for a cassette recorder would include specifications for power input from various sources (batteries, house current, car battery), with specifications for appropriate adaptors. Or a car owner's manual might include specifications for spark plug gap, on the assumption that the owner might do his or her own tune-ups. But a service manual would be much more detailed. Example 7.2 shows the great detail with which a typewriter service manual describes adjustments to the on-off switch.

Information about specifications and tolerances will be found in three contexts in a service manual: in the text of technical background and procedures sections, in tables, and in visuals. (Refer to Chapters 3 and 5 and to the section of this chapter covering visuals for information about the designing of drawings and photos and the proper setup of tables and charts.)

When you include specification information in technical background and procedure sections, you must be especially careful to avoid:

- Letting the numbers get lost in the paragraphs of text
- Letting the numbers obscure the flow of your description of how an assembly works or how a mechanism should be adjusted

This requires constant attention to how numbers relate to the rest of the text. If you have one or two numbers in a long paragraph of explanatory text, the reader can all too easily skim over them and miss what may be vital information. On the other hand, a paragraph loaded with numbers is terribly hard to read. Generally, more than four or five exact numbers in a paragraph of text is too many. The reader simply cannot keep the numbers straight and often loses the line of thought conveyed in the text.

Example 7.2.   Detailed Specifications in Text

1. The Motor Pulley Ring should be positioned at a distance of 2.5–3.0 mm from the end of the Motor shaft.

2. The clearance between the Motor Pulley and the Motor Pulley Washer should be 0.1–0.2 mm.[2]

You may use a number of techniques to solve these problems. To make an occasional numeral stand out in a sea of words, print it in boldface. A manual with different typefaces is a little more expensive to produce but is well worth it if the extra expense ensures that the material is used. As an alternative, the numbers may be separated by white space from the surrounding text. If the number occurs in a procedure description, try to put the exact number in a step of its own rather than including it with other adjustments. Example 7.3 shows how this technique can clarify the text.

If the text contains many exact numbers, it may be better to put them in a separate table or chart. A set of exact numbers can be much more easily assimilated in chart form than in paragraph form. (See Chapter 3, "Organization and Writing Strategies," for more information on tables and charts.) Example 7.4 shows two versions of a description of how to adjust automobile wiper arms, one of which uses a chart to separate verbal from numerical information.

The last technique for making numerical information visible is to include tolerance and specifications in visuals accompanying the text. You must be careful in designing your visuals that numerical information does not clutter up the drawing. A good drawing can easily be ruined with too many labels and excessive specification information, especially in a service manual. Since the audience is generally more technically sophisticated than the audience for an operator manual, the writer may be tempted to

Example 7.3.   Separating Steps in Directions

Original

1. To adjust idle, carefully turn knurled adjustment screw no more than one-quarter turn at a time, until idle speed of 500 rpm is reached.

Improved

1. To adjust idle, turn knurled adjustment screw in or out. Do not turn more than one-quarter turn at a time.

2. Adjust idle to 500 rpm.

Example 7.4.   Using a Chart to Separate Numerical Information

Version 1

With the force applied, the clearance between the tip of the wiper blade and the windshield lower moulding should be ½–2½″ on the right and ¼–2″ on the left for the Dart, and ½–2½″ on the right and ¼–2¼″ on the left for the Coronet and Charger.

Version 2[3] (Actual)

With the force applied, the clearance between the tip of the wiper blade and the windshield lower moulding should be as follows:

Models	Clearance in Inches Between Tip of Blade and Windshield Moulding	
	Right	Left
Dart	½–2½	¼–2
Coronet and Charger	½–2½	¼–2¼

For most readers, the second version will be far more usable.

use unedited engineering drawings for visuals. Although such drawings contain a wealth of specification information, they are usually too cluttered to be useful to the service technician. It is certainly possible to use visuals well to convey specification information, as Figures 7.1 and 7.2 show. Notice especially the use of the close-up to illustrate a particular portion of the drawing. Be sure, if you use visuals to present specification or tolerance information, that the visual is placed right next to the relevant text, particularly in procedures sections.

Model Change Information

Service manuals, unlike operator manuals, are normally updated periodically. The product user does not need update information, since he or she is unlikely to buy successive versions of the same

Figure 7.1. Example of a Visual Used to Convey Specifications and Tolerances. *Source: Single Element Typewriter, Model 200, Service Manual*, Silver-Reed America, Inc. Used by permission.

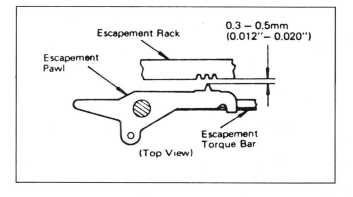

Figure 7.2. Example of a Visual That Conveys Information About Tolerances. Note the effective use of a close-up to avoid cluttering the drawing and to make it more readable. *Source: Single Element Typewriter, Model 200, Service Manual*, Silver-Reed America, Inc. Used by permission.

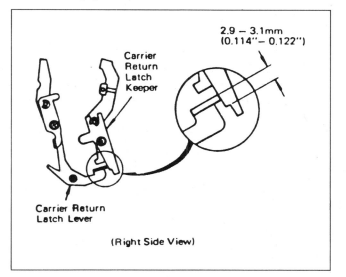

product. In cases where that does happen—a company, for example, may replace office machines on a regular basis—the manufacturer usually supplies a new operator manual with each product. This approach is impractical, however, for the service technician who repairs and maintains a manufacturer's product line over a period of time. One technician, for example, may service several generations of a particular manufacturer's small gasoline engines—as well as those of other manufacturers. Especially when year-to-year modifications are relatively minor, it is much cheaper and handier for the manufacturer to supply supplements to an existing service manual than to write a whole new manual each time the product changes slightly. The writer of a service manual must keep in mind this need for frequent updates and must design the manual so that updating is easily accomplished.

Many different techniques are used to update manuals. Some manufacturers supply replacement or supplemental pages to be inserted into the service manual—which, of course, requires that the service manual be bound in a loose-leaf binder. Increasingly, manufacturers are using video and computer technology to keep their service technicians informed of changes and new procedures. For example, an auto manufacturer may supply its dealerships with a videotape showing a new service procedure being performed, rather than requiring the dealership to send a mechanic to the factory for training. Or technicians may be provided supplemental information on microfiche. No doubt, as the small computer gains widespread use in businesses, new information will be sent out in the form of floppy disks. Although all these technological improvements in communication are desirable, the fact remains that not all users of the service manual have access to the technology to use them. Even if such technology is available, the initial "high-tech" communication should be followed up with "hard copy"—including supplements to the manual. These supplements are usually in the form of individual pages or separate pamphlets.

We recommend the former—replacement or add-on pages—for two reasons:

1. If new information is bound right into the original service manual, the manufacturer is assured that the technician

has the new information. Separate booklets are too easily lost or misfiled.

2. If the new information is bound into the manual at the relevant place, it is more likely to be noticed.

To ensure that update pages are used, the writer of a service manual can use the following techniques:

- At the beginning of the manual, tell the reader that supplements will be provided from time to time, and explain how to use them. Distinguish between add-on pages, which should *follow* existing pages in the book, and replacement pages, which require removal of the old pages.

- Provide a page at the beginning or end of the book on which the technician can record the addition of supplemental material.

- Number pages clearly according to the original manual's pagination, and build in a means of distinguishing add-on from replacement pages. (For example, add-on pages might be numbered with the page number of the page they should follow, plus a letter—26a, 26b, etc.—whereas replacement pages would simply be numbered the same as the pages they are supposed to replace.)

The writer of a service manual should also build in ways to draw the technician's attention to model change information. For example, if add-on or replacement pages are likely to appear in the manual, you may wish to put a reminder at the beginning of each chapter to look for model change information. Or you may wish to have update information printed on stock of a different color from the original page stock.

Field Modifications

You should also include supplemental information about probable field modifications. This information tells the service technician of ways in which the owner may have modified the product. Although these modifications are often not approved by the manufacturer, they do take place, and the technician should be made aware of them. If a particular modification appears often, it may be a signal that a design change is needed in the product— particularly if the modification involves the removal of safety equipment. See Chapter 6 for some discussion of this problem. Sales representatives and local dealers are good sources of information for the service manual writer about what modifications may be expected.

Parts Catalog

A service manual may also contain a parts catalog for the product, although this is often a separate publication. If your manual does include a parts catalog, you must be careful that the principles of good visual design apply. Refer to Chapter 5 for details, but, in general, follow these guidelines:

- Be sure the drawings or photos are clear and large enough to see easily.
- Label each part, preferably with name and order number.
- If you use a system of callouts on the visual combined with a separate list of parts, be sure the list is next to the visual so that the user can refer easily from one to the other.
- If the parts catalog is separate from the manual, be sure that any changes in design that alter the parts catalog are reflected as update information in the manual.

Style

All of the organizational and writing strategies described in Chapter 3 apply to service manuals as well as to operator manuals. The use of such techniques as general-to-specific organization, lists, parallel structure, and active voice are just as important to the reader of the service manual. The differences in style between an operator manual and a service manual have to do primarily with level of language and tone, rather than with the basic principles of presenting information.

The language in a service manual will be a good deal more technical than the language in an operator manual. Since anyone reading a service manual has a certain amount of technical expertise, the writer can use a more specialized vocabulary. Don't make things technical just for the sake of making them technical, however; good writing of any sort is as simple as it can be and still convey the necessary information concisely. Remember also that not all readers of the manual will be familiar with the manufacturer's particular name for things. If you use a term that is special to one manufacturer, try also to include the generic name for the item as well.

The pace of a service manual may also be somewhat faster than that of an operator manual. This simply means that you can present information at a faster rate on the page. You may include less background information and more substantive words per sentence. Remember, however, that the emphasis in a service manual is on procedures, which means that the reader will probably be looking back and forth between the manual and the product as he or she performs a procedure. Keep your sentences and paragraphs relatively short, and use formatting devices to make it easy for the technician to find the right place in the manual again after looking away for a moment to do a step in a procedure. Overloaded sentences and paragraphs are just as annoying to a technically sophisticated reader as they are to a first-time operator.

Finally, the tone of a service manual may be less conversational than that of an operator manual. As we have noted, the function of an operator manual is in part to represent the company to its customers and in part to gently introduce the new user to the product. This requires that an operator manual be written in everyday language and that it sound "friendly." The purpose of a service manual is primarily to explain to a professional or knowledgeable amateur technician how to perform various repair and maintenance procedures. Example 7.5 shows the differences in tone. The example contains two excerpts, the first from a typewriter owner's manual and the second from a typewriter service manual, both dealing with the operation of the right margin stop.

Example 7.5.   Style Differences Between Operator and Service Manuals

> Operator manual:
>
> The right margin stop prevents you from typing past the right margin; however, you can space or tab right through it. To type past the right margin, press MAR REL (margin release) and continue typing.[4]
>
> Service manual:
>
> As the Carrier moves to the right still more after the Bell ringing, the Margin Stop Latch moves up the Margin Stop Right extension allowing the Margin Rack to rotate. Then the Margin Plate attached to the Margin Rack rotates the Linelock Bellcrank through the Margin Link. At that time, the Linelock Bellcrank extension moves the Linelock Keylever downwards causing its extension to insert into the space between the Keyboard Lock Balls. And the Keyboard has been locked to prevent typing past the Margin Stop Right.[5]

To sum up, the stylistic principles involved in writing operator manuals and service manuals are the same; the principles are applied differently, however, as a result of the differences in audience.

Visuals

As with the verbal portion of a service manual, the basic principles for visual design (explained in Chapter 5) apply to both operator manuals and service manuals. Good visual design is perhaps even more important in service manuals because so much of a service manual is devoted to procedures for repair and adjustment. The combination of verbal and visual material must make the procedure perfectly clear. This often means that the balance of material shifts toward the visual: a service manual will tend to have more drawings and photographs than an operator manual.

Although the principles are the same for both kinds of manuals, again their application differs. A service manual will have more technical drawings; exploded diagrams and cutaways rather than perspective drawings; and circuit diagrams rather than block diagrams. You must take great care to ensure that these are large enough to see easily and are not cluttered.

An exploded diagram, for example, can often be made much more comprehensible by dividing it into sections. Figure 7.3 shows how "sectionalizing" an exploded drawing of a transmission permits more complicated portions to appear in close-up. The whole view could have been laid out in one piece and photo-reduced to fit on the page, but the result would have been difficult to read.

As suggested in Chapter 5, when the complexity of the drawing permits, label parts with the part name rather than a callout. Since the technician using the manual will already be looking back and forth between the manual and the product, adding another place to look (the key that identifies the callout) will only increase the possibility of a mistake. You must also ensure that the lines showing how parts fit together are easily distinguished from lines or arrows leading from labels or callouts. One good way to do this is to use broken lines for the former and solid, heavier lines for the latter.

If you use cutaway drawings, be sure that the reader can easily differentiate the "layers" of the cutaway. Often you can do this by careful shading—but be careful that your shading does not clutter the drawing. At other times, the best choice may be to use color to highlight different levels. Figure 7.4 shows how a photograph (or drawing made to simulate a photograph) can be used as a cutaway. Notice how easy it is to distinguish the parts.

Perhaps the most easily abused form of illustration is the circuit diagram. The writer of a service manual should avoid the temptation to use the schematic developed with the product. First of all, the original schematic was probably drawn on a large scale. Reducing it to fit onto the manual page would render it unreadable. Second, it may contain more detail than the technician needs, which may lead to unnecessary clutter. Instead, have a schematic drawn for the manual, one that includes only necessary information and is drawn to a scale appropriate for the manual page size.

Figure 7.3.   Example of "Sectionalizing" an Exploded View to Make It Appear Less Cluttered and to Enable the Reader to View Smaller Parts in Close-up. *Source: Dodge Dart, Coronet and Charger Service Manual 1967,* Dodge Division, Chrysler Motors Corporation. Used by permission.

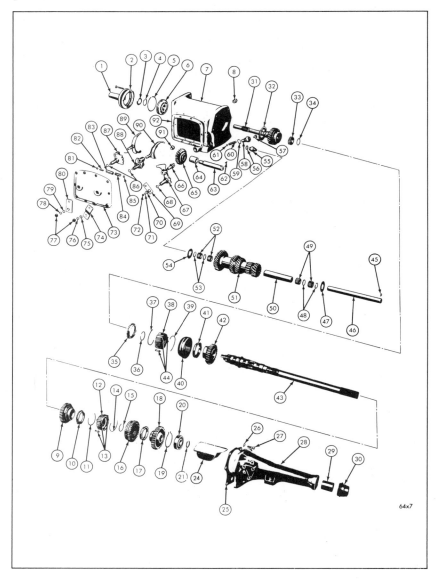

Figure 7.4.   Example of a Photograph Cutaway. Note how easy it is to distinguish the different parts because of the photographic appearance. *Source: Dodge Dart, Coronet and Charger Service Manual 1967,* Dodge Division, Chrysler Motors Corporation. Used by permission.

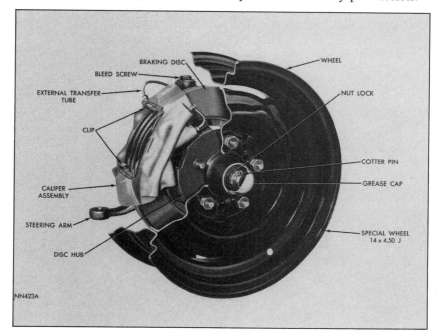

If your product's manual includes a circuit diagram, you must be sure that your readers can interpret the symbols used. For a device that is primarily electronic, this is not usually a problem. Someone without knowledge of electrical circuitry is not likely to use a radio service manual, for instance. However, for a product in which an electrical system is only one component—farm equipment, for example—and for which the service manual's users are likely to be diverse, you may wish to include additional information. For example, the writers of a tractor service manual included the chart shown in Figure 7.5 as explanation for electrical symbols.

Good visual design simply means making sure that your visuals are big enough to be easily seen, are as simple as they can

Figure 7.5.   Page of Electrical Symbols Included in a Manual to Help Readers Find Their Way Through Circuit Diagrams. An excellent example of writing with the user in mind. *Source: Service Manual, Series 2 Four-Wheel Drive Tractors*, applicability: 1977 Production, Versatile Farm Equipment Company, a division of Versatile Corporation. Used by permission.

SYMBOL	MEANING	SYMBOL	MEANING
	Wires crossing. No connection.		Mechanically actuated switch: normally closed, held open.
	Wires connected.		Multi-position rotary switch. Connections and positions as tabulated in diagram.
	Ground connection		
	Meter or gauge, as labelled.		Relay, single pole.
	Motor, DC.		
	Lamp, illuminating.		Solenoid and Valve
	Lamp, indicating.		
	Battery: two or more cells.		Resistor, fixed.
	Switch, general.		Resistor, variable.
	Pushbutton, normally open.		Fuse, current rating as labelled.
	Pushbutton, normally closed.		Circuit breaker, current rating as labelled.
	Thermostat switch, closes on rising temperature.		Compressor clutch.
	Pressure sensor, closes on rising pressure.		Speaker.
	Pressure sensor, opens on rising pressure.		

be while still conveying the necessary information, and are clear enough to be easily understood. How these basic principles are put into practice depends on your audience. Since the audience for a service manual is likely to be more knowledgeable than the audience for an operator manual, you can use more technically sophisticated visuals—but you should still make them big, simple, and clear.

Format and Mechanics

Format

Service manuals differ from operator manuals in their large-scale organization and in the relative importance of certain large-scale features. The organization is still determined by "user questions," but the users of service manuals ask different questions. The new owner of a product is concerned with how the product works, how to care for it, and so on. The service technician wants to know how to service or fix the product. The new owner will probably flip through the pages of the operator manual, reading more or less at random. The service technician will look for a specific section that covers the necessary procedure and read that section only, unless specifically directed elsewhere. Thus, a service manual must be organized to help the reader locate the procedure or explanation needed for a particular job and to direct the reader to other relevant sections.

We noted in Chapter 2 that most operator manuals are organized by general categories of information, such as start-up procedures, maintenance, storage, and so on. Generally service manuals are organized by product system: for example, an auto service manual will have chapters on the engine, the transmission, the cooling system, the electrical system, and so on. This kind of organization makes finding the information needed quite simple. If the car has a problem with the cooling system, the technician knows exactly which chapter to read. Often each chapter will have its own table of contents, as in Figure 7.6.

This organizational system works very well when the technician knows the precise location of the problem. However, the systems of a product interact, and the cause of the malfunction may not be immediately obvious. Because the service manual divided up by sections does not so easily show the overlap of systems, the writer must take care to include a comprehensive troubleshooting section that directs the reader to the appropriate pages on the basis of symptoms rather than identification of the problem. In other words, instead of listing only "carburetor adjustment, ... p. 3–23," it would also be helpful to include "car hesitates .... see carburetor adjustment, ... p. 3–23."

Figure 7.6. Sample Table of Contents for a Chapter in a Manual. This kind of sectionalizing keeps the main table of contents from becoming too cluttered and is another example of general-to-specific organization. *Source: Service Manual, Series 2 Four-Wheel Drive Tractors*, applicability: 1977 Production, Versatile Farm Equipment Company, a division of Versatile Corporation. Used by permission.

SECTION 2: ENGINE SYSTEMS

Table of Contents

For the same reason, the writer of a service manual must include a comprehensive index. If possible, include cross-referencing within the index: for example, "carburetor, adjustment (see also fuel filter)." The idea is simply to make the manual useful to the technician by making it as easy as possible for him or her to find the section needed.

The problem of how to organize the information in a service manual becomes particularly difficult when the manufacturer decides to combine the service manual and the owner's manual in one book. Although one book is cheaper to produce than two and ensures that everyone has the same information, we do not recommend this practice. As this chapter has shown, the service manual has a very different audience and purpose than the operator manual, and this difference should be reflected in content, style, and format. To combine both kinds of manuals in one book makes it nearly impossible to maintain the appropriate distinctions. Furthermore, including the service manual with the operator manual may encourage some owners to perform procedures they should not perform—because they aren't skilled enough or don't have the appropriate tools. We believe that it is much better for the manufacturer to keep operator manuals and service manuals separate: if the owner *is* a skilled amateur, he or she can always write the company and order the service manual.

Mechanics

Many of the same considerations apply to the mechanics of a service manual as apply to operator manuals, but some differences should be noted. First, a service manual will probably get much harder use than an operator manual. Once an owner has learned how to use and care for the product, the operator manual will probably lie untouched unless a problem occurs. The service manual, in contrast, will be used day in and day out. Even if the technician is thoroughly familiar with a particular procedure, he or she will still need to check information on tolerances and specifications. The cover must stand up to this hard use; usually plastic or (at least) coated stock is required for a service manual cover.

We have discussed the importance of making the manual easy to update. For this purpose, we recommend some kind of loose-leaf binding—for example, a three-ring binding, a clamp-type binding, or a combination of the two. If you use a three-ring binding, be sure that the pages are heavy enough or reinforced so that they will not tear out under hard use. In general, the pages of a service manual must be heavier and more soil-resistant than those of an operator manual simply because of the harder use expected.

Finally, the service manual is usually bigger than an operator manual: a standard page size is 8½ by 11 inches. Since a service manual is usually not used out of the shop, it need not be as portable, and the larger page size makes the drawings easier to read.

Summary

We have seen in this chapter that, although the same basic principles apply to the writing of both operator manuals and service manuals, the application of those principles differs because the two kinds of manuals have different audiences and different purposes. Although both manuals may be written about the same product, they will differ in content, style, and organization. The service manual is written for an audience that is more technically sophisticated and is interested primarily in procedures for service and repair. Therefore, the manual will contain information about the technical background of a product system and about procedures for repair. Since the audience is more technically sophisticated, more technical language and a less conversational style may be used. Because of its specialized purpose, a service manual will be organized to assist the technician in finding the exact repair procedure needed. All the differences notwithstanding, the design procedure for both types of manuals is the same: clearly define the audience and purpose, and then arrange and write the material to reflect that definition.

References

1. *Volkswagen Rabbit/Scirocco Service Manual, 1980 and 1981 Gasoline Models Including Pickup Truck 1981,* Cambridge, MA: Robert Bentley, 1981, p. 61. Used by permission.

2. *Single Element Typewriter, Model 200, Service Manual.* Torrance, CA: Silver-Reed America, p. 5. Used by permission.

3. This is the actual version, taken from the *Dodge Dart, Coronet and Charger Service Manual 1967.* Detroit, MI: Dodge Division, Chrysler Motors Corporation, 1967, p. 8-89. Used by permission.

4. Reprinted by permission from *IBM Correcting Selectric II Operating Instructions.* © 1973, p. 6.

5. *Single Element Typewriter, Model 200, Service Manual.* Torrance, CA: Silver-Reed America, p. 33. Used by permission.

8

Manuals for International Markets

Overview

In the last two decades, international trade has grown in volume and complexity. If you have attended national conferences and trade fairs, you are doubtless aware that "Think international" has become a familiar slogan and that the economic links between nations are steadily increasing. As a consequence, even small and intermediate-size companies whose markets have historically been confined to the United States now find themselves selling computers in Africa, rice planters in Southeast Asia, trucks in China.

When a company decides to market its products outside the United States, its manuals, along with its service and marketing publications, may have to be produced in languages other than English. Because English is the predominant language of international trade, competent English-speaking representatives or translators will usually be on hand at initial negotiation and contract stages. However, when products actually begin to be sold and used in other countries, written materials in the native languages become a necessity. Those companies already involved in international trade expect to produce their manuals in some or all of the following languages:

Afrikaans	Hungarian
Arabic	Indonesian
Danish	Italian

Dutch	Japanese
English	Norwegian
Farsi	Portuguese
Finnish	Serbo-Croatian
French	Spanish
German	Swedish
Greek	Turkish
Hebrew	

In this chapter we will discuss some of the special problems of manuals in translation and suggest ways to make the translation job easier and more cost-effective.

North-South Markets and Manual Users

If international marketing is a new venture for your company, some general observations about the probable characteristics of your manual users may be of help. The Brandt Commission, in 1980, published an economic study that categorized industrialized and developing nations in a new way by dividing them into North and South. This rough geographical division provides a convenient way to think about differences in users of your manual. The industrialized North includes North America, Europe, Russia, Japan, and anomalous southern "pockets" such as Brazil, Australia, and New Zealand. The developing nations of the South include most of South and Central America, Africa, and Southeast Asia. (Trade with China is a special case, since China only recently entered the world trade market and has not yet clearly identified or aligned itself in the world economic order.)

Characteristics of the North: Industrialized Nations

The GNP and per-capita income of the North is higher, its technology more advanced, and its consumer demands for American products more specialized. Products most in demand in the North are those domestically in short supply, domestically expensive, or preferred because the American technology of the product is considered superior. Manual writers can thus make the following assumptions about manual users in industrialized nations:

- Literacy levels will be high.
- Users will probably be familiar with standard technologies.
- Service and repair support systems will be established and available.

Characteristics of the South: Developing Nations

The GNP and per-capita income of developing countries will be low, the technologies scattered and unreliable, and the demand for imported consumer goods more diverse—everything from clothing, cosmetics, and baby food to appliances, automotive products, plumbing supplies, and medical equipment. Manual writers can thus make the following assumptions about manual users:

- Literacy levels may be low, and illiteracy more prevalent.
- Users may be unfamiliar with technologies, even the simple technologies.
- Service and repair support systems may be spotty, primitive, or nonexistent.

Problems in Producing Translated Manuals

The chief problem of producing good translated manuals is their expense. Translated manuals will be the most expensive per copy of all your publications. Ideally, if your manual needs translation into one or several languages, you should try to create an English original manual that needs a minimum of changes. Companies already involved in multiple language manual production identify these as key problems:

1. Translated manuals are short runs, and the fewer copies of small-batch runs mean higher costs per copy.
2. Labeling on visuals must be redone.
3. Nomenclature for certain parts and tools varies from language to language.

4. Service and repair systems vary widely in quality and comprehensiveness from country to country.

5. Parts identification and replacement become more difficult as the supply line lengthens (the more remote the country and the poorer its infrastructure of roads, railway, and air travel, the worse the problem).

6. Cultural differences may affect manual use. For example, some countries have culturally entrenched prohibitions against performing certain mechanical and maintenance tasks. Rural areas, in particular, may have scant understanding or familiarity with machines and their uses.

Let's take a look at these problems in more depth.

Translation

Manuals meant for distribution in industrialized countries are obviously easier to prepare. At the end of this chapter we list some sources for translation of technical materials into the standard Romance and Germanic languages. French, Spanish, Italian, and German are often sufficient for European markets. If your manuals make use of a well-developed and standardized vocabulary, computer translation into French, Spanish, and German is now possible to about 80 percent accuracy.

Manuals produced in the more "exotic" languages are harder to deal with. Many companies have found that their own dealers and international representatives can serve as translators or can identify competent translators within the country. The use of company dealers and service representatives as translators is especially valuable because they know the products as well as the language.

If you are marketing products in Canada, remember that Canada is bilingual. Canadian Law 101 now requires that consumer publications (i.e., those expected to be used by the general public), such as operator manuals, be printed in both French and English. English suffices for the more technical publications, such as service manuals, to be used by technicians only.

If you are marketing products in Mexico, labeling should, of course, be in Spanish. Bilingual labeling is desirable in the United States for a number of products, especially those used in agriculture (farm machinery, pesticides, herbicides, fertilizers) where the labor force may be Spanish-speaking. Your labels and warnings on such toxic or dangerous products should be in Spanish as well as English.

Sometimes companies are able to identify particular ethnic groups as closely associated with a particular trade or industry. For example, one American manufacturer of a paint sprayer that, when misused, could be dangerous, also recognized that a large percentage of its customers were Greek-speaking contract painters. That company now prints some of its instructions and danger warnings in Greek as well as English.

Visuals

The most important component of manuals in translation is the visual. When visuals are clear and self-explanatory, they also help to diminish whatever errors or mistranslations might creep into the translated verbal text.

All of the suggestions for good photos, drawings, and charts suggested in Chapters 3 and 5 hold true for translated manuals. Visuals should be big, simple, and clear. They should be carefully coordinated with verbal text and planned so that the most important elements are visible.

Visuals are particularly crucial when user literacy levels are likely to be low. If you find that, on an average, you are devoting less than half of your manual to visuals, try to add more illustrations. Do this even if, to you, the mechanism or procedure seems perfectly obvious. We recommend, too, that you pay special attention to safety labels and warnings expressed in words alone. The addition of pictograms or drawings to warnings may well make the difference between a user's recognizing a danger and missing it entirely.

You can grasp the importance of visuals by imagining your own reaction to a manual originally written in Chinese and poorly translated into English. Your lifeline to assembly, use, and maintenance of the product will be the photos and drawings.

We suggested in Chapter 5, "Visuals," that you use labels

directly on the visual whenever possible. Manuals for translation are a notable exception, largely because of costs. Key numbers and accompanying legend in English can easily be translated by changing the legend only.

For visuals in translated manuals:

- Follow suggestions for visual effectiveness in Chapter 5.
- Add more visuals if verbal text exceeds visuals in space allotment.
- Use key numbers and legend to simplify translation of complex visuals.
- If possible, incorporate visuals or pictograms into safety warnings and labels.

Nomenclature

Names for parts, processes, and procedures vary from language to language. Sometimes commonly used English words have no real equivalents. In fact, even "English" and "American" differ notably. For example:

English	American
petrol	gas
earth (electrical)	ground
flex	wire
spanner (generic)	wrench
bonnet (auto)	hood
boot (auto)	trunk
dynamo	generator

What may be a "shovel" in English may be a "spoon" in another language.

Many "mature" industries, those that for decades have been producing products whose essential features change little from year to year (e.g., plows, automobiles, typewriters), have created comprehensive glossaries or dictionaries of routine industry terms. These terms are common parlance, that is, everyone understands the difference between a knob and a handle and knows what is meant by a socket wrench or a tire iron.

In translated manuals, you should make an effort to standardize the vocabulary for parts and processes and to be consistent in describing routine procedures. For example, do not instruct the user to "oil" the machine at one time and to "lubricate" it the next, or to "monitor the needle for pressure variations" the first time and to "check the gauge" the second. (Note that the terms *oil* and *lubricate*, as well as the terms *monitor* and *check*, actually have slightly different meanings in English but are sometimes used loosely or interchangeably.)

Standardized vocabulary is also important in processes and procedures. The following are some commonly used procedural verbs:

tighten	remove	raise	press
loosen	add	lower	release
fill	check	attach	stop
empty	place	fasten	start
clean	turn	adjust	replace

Make decisions about the procedural words you will need to use, and stick to them. Don't, for example, say "lower the arm" in one place and "allow the arm to drop" in another, or "turn the wheel to the right" and then "rotate the wheel clockwise."

Figure 8.1 shows two pages of a translated manual, one page the English version, the other the Finnish version. Notice these features:

- Vocabulary is standardized (choke, valve, stop, start).
- Chart layout makes reading easier.
- Symbol, symbol name, and symbol meaning are clearly laid out and reproduced identically from one language to the other.
- English version is on page 3; Finnish version is also on page 3 of the Finnish section.

Service, Repair, and Parts Replacement

Americans are accustomed to convenient and readily available service and parts replacement. Remember the frustration in the early years of foreign cars in the United States when customers complained about waiting weeks for a part? In international trade to industrialized countries, the advent of modern inventory control and computer update has produced vast improvement in the delivery system of service and repair.

Visuals are again of crucial importance in translated service manuals and parts lists. Many sophisticated and advanced systems of manual production have developed parts catalogs and repair manuals that consist almost entirely of visuals and numbers. Parts lists contain only an identifying picture or drawing and an identification number, and service manuals are done almost entirely with photos of the product, photos of the tools needed to assemble and maintain the product, and a limited "universal" vocabulary.

Cultural Differences

When new or unfamiliar products are introduced, particularly in developing countries, cultural patterns may significantly affect their use. Ask international representatives about their experiences in the field, and they will quickly regale you with anecdotes. You will hear about eggs fried on electric irons, refrigerators used as air conditioners (just leave the door open), wood fires built in the cavities of gas ovens, tires mounted backwards on rims. One frustrated farm equipment representative found that the only way he could convince illiterate peasants that a chopper was a dangerous machine was to toss a hapless farmyard cat into the mechanism.

Such anecdotes reveal difficulties to be found in cultures where technologies are unfamiliar. Developing countries will present many such instances. In addition, the social structure of these countries may be much more rigidly hierarchical, with clear divisions of labor among classes or castes of people. Thus, one worker may be allowed to drive a vehicle but not to change a tire, and fixing a machine may be considered demeaning or socially taboo.

In brief, if your product is being marketed in a developing country whose culture is markedly different from yours, then your manuals must be crystal clear and as simple and graphic as possible.

Figure 8.1.   Translated Manual with Standardized Vocabulary and Symbol Explanation (English and Finnish). The chart layout and one-word system for symbol names plus symbol explanation make translation easier. *Source: Evinrude® Outboards*, 1983, Outboard Marine Corporation. Used by permission.

Symbol	Symbol Name	Meaning or Purpose of Symbol
"Functional Description" Symbols		
	CHOKE	Identifies CHOKE control.
	VALVE	Identifies a control valve.
	STOP	Identifies STOP SWITCH control. May also identify STOP position of throttle control on certain motors.
	START	Identifies position of throttle control device during starting. May also identify STARTING control.
	START-MOTOR	Operating device for starting motor.
"Instructional" Symbols		
	LATCH	Identifies device provided to LATCH or UNLATCH engine cover.
	FUEL SHUT OFF	Identifies device provided to cut off fuel supply to engine.
	SPARK ADVANCE	Number (in degrees) following this symbol indicates recommended maximum spark advance for engine. (Symbol and number appear on engine surface.)
	KEROSENE	Indicates KEROSENE is to be used or identifies KEROSENE is present.
	.FUEL	Indicates GASOLINE is to be used or identifies GASOLINE is present.
	OIL	Indicates OIL is to be used or identifies OIL is present.
50/1	FUEL OIL MIX	Identifies FUEL/OIL mixture for 2-stroke engine. Indicates each 50 parts of gasoline are to be mixed with 1 part of oil. Mixture to be mixed completely.

3

Figure 8.1. *(continued)*

VERTAUSKUVA	VERTAUSKUVAN NIMI	VERTAUSKUVAN TARKOITUS TAI VAIKUTUS
"TOIMINTASELOSTUS" VERTAUSKUVAT		
	KAASUTIN	Osoittaa kaasuttimen saato.
	VENTTIILI	Osoittaa saatoventtiili.
	PYSAHDYS	Osoittaa pysahdyskytkin. Voi myoskin osoittaa kaasuvivun pysahdysasento.
	KAYNNISTYS	Osoittaa kaasuvivun asento kaynnistaessa. Voi myoskin osoittaa kaynnistyssaato.
	MOOTTORIN KAYNNISTUS	Moottorin kaynnistyslaite.
"OHJEITA ANTAVIA" VERTAUSKUVIA		
	SALPA	Osoittaa laitteen joka lukitsee tai irroittaa moottorin kannen.
	POLTTOAINEKAT-KAISU	Osoittaa laitteen joka sulkee moottorin polttoainesyoton.
	SYTYTYKSEN SAADIN	Numerot (asteissa) jotka seuraavat tata vertauskuvaa suosittelevat maksimaalinen sytytyksen saadin moottorille. Vertauskuva ja numerot ovat moottorin pinnassa.
	PALOOJY	Nayttaa etta palooljya on kaytettava tai identifioi etta polttooljy on olemassa.
	POLTTOAINE	Nayttaa etta bensiinia on kaytettava tai identifioi etta bensiini on olemassa.
	OLJY	Nayttaa etta oljya on kaytettava tai identifioi etta oljya on olemassa.
	POLTTOAINE-SEKOITUS	Osoittaa polttoaine/oljy sekoitus 2-tahti-moottorille. Nayttaa etta joka 50 bensiinin osa pitaa sekoittaa 1 osalla oljya. Sekoitus on oltava taysin sekoitettu.

3

A final word: many underdeveloped cultures are remarkably adaptive to new technologies and products, once their people understand the underlying principles of the technology. For example, the technological adaptiveness and dexterity of the Eskimos are legendary. Once they understood that fuel is to an engine what fish is to a sled dog, and that a snowmobile was not "dead" when the first tank of gas was gone, they were soon zipping over the ice, skilled in the operation of snowmobiles.

Packaging the Translated Manual

Translated manuals are frequently packaged in parallel columns, with English on one side, and the translation on the other, or with a "shared visual" format, as shown in Figure 8.2.

Figure 8.2.   Sample Layout of Multilingual Manual, Showing "Shared" Format. The two sample pages in this figure show typical multilingual layouts. The prose text is often laid out in parallel columns (e.g., English and French), which allow you to use single photos or visuals that are shared by the various columns of text.

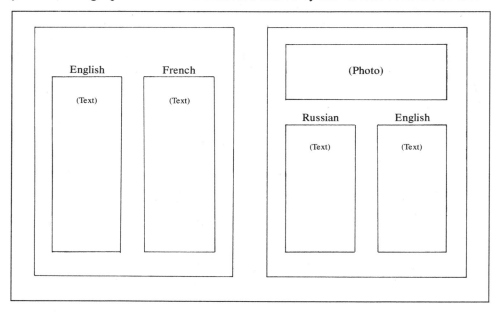

When a manual is relatively short and simple, you can make one manual serve for three or four languages by experimenting with foldouts and shared visuals. For example, you can devote the left page to a photo or drawing of the product and the right page to a foldout with French, German, and/or Spanish texts laid out identically and keyed identically to the shared visual. Figure 8.3 shows how a short manual for translation can be formatted to ease translation and keep costs down. The English version of the manual is 18 pages long. The international manual has the following features:

- One international manual, bound as a single book, serves for eight languages (English, German, Italian, Spanish, Dutch, French, Norwegian, Swedish, Danish, Finnish).

- Each language has a separate section (i.e., a French section, a German section, etc.).

- Page layout, number of pages, and page numbering system are identical for each section.

- Shared visuals can be opened out to use with any of the language sections because legends and callout numbers are identical.

On the following pages, we show you examples of an English page and its Italian equivalent (see Fig. 8.3). You will also find addresses of some of the major technical translating services.

Summary

Whenever manuals must be translated, remember these two guidelines:

1. Standardize your vocabulary.
2. Show rather than tell about product and procedures.

Figure 8.3.   Translated Manual with Shared, Foldout Visual and Stand-ardized Format and Vocabulary. Note the following features: layout for Italian and English versions are identical (as are the other six language sections included in this manual); page numbering systems are identical (makes cross-referencing from one language to another easier); the

(A)	**(B)**	**(C)**
Incorrect	**Incorrect**	**Correct**
OVERLOAD FORWARD CAUSES BOAT TO "PLOW"	OVERLOAD AFT CAUSES BOAT TO "SQUAT"	BALANCED LOAD GIVES MAXIMUM PERFORMANCE

⚠ **Safety Warning: If engine is tilted forward so as to cause plowing (see A), swamping may occur in rough water. If engine is tilted aft so as to cause porpoising (see B), steering may be erratic or unstable. See correct angle adjustment (see C).**

Lubrication

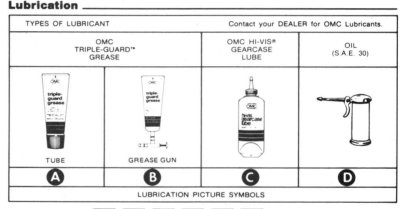

TYPES OF LUBRICANT		Contact your DEALER for OMC Lubricants.
OMC TRIPLE-GUARD™ GREASE	OMC HI-VIS® GEARCASE LUBE	OIL (S.A.E. 30)
TUBE GREASE GUN		
A **B**	**C**	**D**
LUBRICATION PICTURE SYMBOLS		

LUBRICATION POINTS 30 31 32 33 34 35

30. GEARCASE LUBRICATION
   Remove oil drain/fill and oil level plugs from side of gearcase. With motor in normal running position, allow oil to drain completely.
   To refill, place tube of OMC HI-VIS® Gearcase Lube or equivalent in drain/fill hole. If OMC HI-VIS Gearcase Lube is not available, OMC Premium Blend Gearcase Lubeor equivalent can be used as an alternate. With motor in normal running position, fill until lubricant appears at oil level hole. See **Specifications** for gearcase capacity.
   Install oil level plug before removing lubricant tube from oil drain/fill hole. Drain/fill plug can then be securely installed without oil loss.
   If the proper tube or filler type can is not available, install drain/fill plug. Slowly fill gearcase through oil level hole allowing trapped air to escape. Install plug.

A. Oil Level Plug
B. Oil Drain/Fill Plug

Change after first 20 hours of operation and check after 50 hours of operation.
Add lubricant if necessary.
Drain and refill every 100 hours of operation or once each season whichever occurs first.

Note	Note: Recommended lubricants which have been formulated to protect against damage to bearings and gears must be used as extensive damage can result from improper lubrication.

31. Idle Speed Adjusting Knob Shaft, Spark Advance Linkage, Cam Roller, Shaft and Gears
32. Swivel Bracket, Engine Cover Latch Shaft
33. Shift Lever Shaft and Detent, Choke and Carburetor Linkage
34. Clamp Screws, Tilt/Run Lever Shaft, Tilt Shaft, Steering Handle, Throttle Shaft and Gears
35. Steering Handle Throttle Gear and Bushing

Frequency of Lubrication	
TYPE OF USE	FREQUENCY
Fresh water	Every 60 days
Salt water	Every 30 days
Storage of 30 days or longer	Before placing in storage

Figure 8.3. *(continued)*
numbers in boxes (30, 31, 32, 33, 34, 35) are lubrication points and correspond to numbers 30–35 on the foldout shared visual. *Source*: *Evinrude® Outboards*, 1983, Outboard Marine Corporation. Used by permission.

⚠ Avvertimento di Pericolo: Se il motore è inclinato troppo in avanti, la prua si affossa nell'onda e si rischia di limbarcare acqua. Se il motore è inclinato troppo indietro. La barca picchia e la guida diviene incerta od instabile. Cfr. corretta regolazione dell'angolo.

Lubrificazione

TIPO DI LUBRIFICANTE		Rivolgersi alla CONCESSIONARIA per i lubrificanti OMC	
OMC™ TRIPLE-GUARD GREASE		"HI-VIS"® GEARCASE LUBE OMC	OIL (S.A.E. 30)
TUBO	SIRINGA DI GRASSAGGIO		
A	**B**	**C**	**D**
SIMBOLI GRAFICI PER LA LUBRIFICAZIONE			

PUNTI DA LUBRIFICARE

30. PER SCARICARE LA SCATOLA INGRANAGGI

Togliere i tappi filettati di scarico/rifornimento e di livello a lato della scatola ingranaggi. Con il motore in normale posizione di corsa, lasciar defluire tutto l'olio.

Per rifornire, imboccare il tubo di "HI-VIS® Gearcase Lube" OMC od equivalente nel foro di scarico/rifornimento. Se il lubrificante "HI-VIS Gearcase Lube" OMC non fosse reperibile, si potrà ripiegare sul "Premium Blend Gearcase Lube" OMC o suo equivalente. Con il motore sempre nella normale posizione di corsa, riempire finchè il lubrificante sale a lambire il foro di livello. Per la capacità della scatola ingranaggi, cfr. **Specifiche.**

Montare il tappo di livello dell'olio prima di staccare il tubo del lubrificante dal foro di scarico/rifornimento. Si potrà così riavvitare il tappo di scarico/rifornimento senza perdita d'olio.

Se non si dispone del tubo o della siringa adatti, montare il tappo di scarico/rifornimento. Riempire lentamente attraverso il foro di livello, permettendo all'aria di fuoriuscire. Riavvitare il tappo.

A. Tappo livello olio
B. Tappo per scarico/riempimento olio

Cambiate l'olio dopo le prime 20 ore di funzionamento, quindi verificare il livello ogni 50 ore. Se necessario, rabboccate.
Cambiate l'olio dopo ogni 100 ore di funzionamento o, comunque ad ogni stagione.

Note Nota: Bisogna usare i lubrificanti consigliati che sono stati formulati per la protezione dei cuscinetti e degli ingranaggi dato che l'uso di un lubrificante non adatto può arrecare danni notevoli.

31. Per il minimo regolare il pomello dell'asse, l'anticipo di accensione la camma cilindrica, l'asse e gli ingranaggi
32. Cavalletto di brandeggio ed alberino della leva di fissaggio della carenatura del motore
33. Alberino e dente di arresto della leva comando cambio; articolazione dello starter (comando gas) e del carburatore
34. Viti di chiusura, ingranare e disingranare la leva dell'asse, disingnare l'asse, la leva di direzione, l'asse e i comandi della farfalla
35. Ingranaggio comando gas e cuscinetto della leva di direzione

Frequenza di Lubrificazione	
TIPO DI UTILIZZAZIONE	FREQUENZA
Acqua dolce	Ogni 60 giorni
Acqua di mare	Ogni 30 giorni
Rimessaggio di almeno 30 giorni	Prima del rimessaggio

Addresses of Translation Services

If translating your own manuals is beyond your expertise, you may be able to use a translating service. The following are addresses of some of the major ones:

AdPro Associates, Ltd.
P.O. Box 383
Cedar Rapids, IA 52406

Bilingual Productions, Ltd.
164 South Service Road
P.O. Box 94
Stony Creek, Ontario L8G 3X7
Canada

INTRIN
820 Valley Forge Plaza
King of Prussia, PA
19406

ITA
4010 Washington
Kansas City, MO 64111

SH[3]
408 West 83 Terrace
Kansas City, MO 64114

9

Managing and Supervising Manual Production

Overview

Manual production cannot be treated in isolation from the company structures that surround it. This chapter is written to help those who are setting up a manual production operation for the first time and those who realize that their present setups do not seem to be working as well as they should.

The quality of service publications depends on a number of key factors: the initial choice and training of writers, the structure and managerial philosophy of the company, the clear delineation of lines of authority, and the fulfillment of writers' basic needs. We include this chapter because we know that the techniques and suggestions we have made to manual writers cannot be effectively applied unless the fundamental company structures are well designed. Good management exists to make it possible for people to do their jobs well.

Who Writes the Manual?

Before we begin a seminar on manual writing, we analyze our participants by asking them to fill out a personal information sheet listing their experience and training for manual writing. Here are just a few of the answers:

Engineering graduate (all kinds)

Service and parts manual writer

English or journalism graduate

Law, business, or psychology graduate

Technician

Prototype builder

Product safety manager

Company owner

Son of company owner

Transfer from marketing or advertising

Magazine science writer

No experience

Clearly, the entry into specialized technical writing is sometimes through the "front door," but far more often it is an outgrowth of other job duties, a discovery or tapping of a special talent in mid-career, a tangential assignment, or a deliberate second career choice. In small and intermediate-size companies, manual writing is often an add-on to many other job tasks, and writers may be given little guidance in how to proceed.

Choosing the Technical Writer

When a company decides to assign the manual-writing task or to hire a new writer, it often asks, "Should we choose technicians and engineers and then teach them how to write, or should we choose professional writers and teach them the technology of the product?" Posing the question this way can be misleading because of the underlying assumptions—that technicians and engineers can't (or won't) write and that trained writers will probably be technically naive or ignorant.

A better way to think about the choice is to choose someone who *can* communicate and *likes* to. Certainly, some lawyers and engineers are so bound up in jargon that they find it almost impossible to simplify a message for a general public audience, and some wordsmiths write clean and explicit prose yet are so technically inept that they cannot grasp the workings of the simplest machine. Nevertheless, there are individuals who possess a combination of the necessary communication skills, and these make the best technical writers. They have technical sense about how things work and either know the product from experience elsewhere in the company or can, with a minimum of explanation and hands-on practice, quickly grasp a new technology or product. They recognize clear, correct prose and can also write it. They have a visual sense about drawings, photos, and format devices.

As a manager, you can devise appropriate screening devices to identify good communicators. One of the most effective is to give applicants a simple device or product, along with basic information and relevant photos or drawings. Then ask them to write and lay out a sample page of a manual. Another method is to give applicants sample pages of manuals. Choose a spectrum (some good, some bad, some average), and ask applicants to rank the pages and to justify, in writing, the reasons for their choices.

Writer Training

Writers need training for manual production, just as they do for any other kind of specialized publication. We have found that some companies provide no training whatsoever or give their writers only the sketchiest of orientations, whereas others have systematic and comprehensive training programs.

The best training programs are those provided before manual writing begins, as part of the planning process. After orientation in the fundamentals of manual production, writers are then given periodic training in special skills. In the long run, providing preliminary training is more cost-effective than waiting until writers are floundering. It may then be too late to correct errors, after considerable money has already been spent on such items as art, photography, or printing.

Companies that have been producing products for a long time, especially if they are also large companies, often have sophisticated and comprehensive training programs for manual writers. If you work for such a company, you may already have had orientation sessions, hands-on practice working with other, more experienced writers, and close contact with your publication managers and editors. If your company has no such training program, here are some of the training techniques you might consider:

- Have new writers work through a manual from start to finish with an experienced writer.
- Give new writers in-house handbooks and style guidelines or workbooks to orient them to company procedures.
- Have managing editors work closely with new personnel in the first months on the job.
- When companies are decentralized and manuals are produced at several places, assign one manager to coordinate quality control of the manuals (Deere and Company, for example, has a "roving editor" who travels among the various manual production locations).
- Bring writers together for yearly training sessions on such special work topics as metric conversion, the writing of safety warnings, or photographic techniques.
- Make libraries and files of company and competitors' manuals available for writers to look at.
- Teach writers "incrementally" by assigning only small segments of a manual for their first assignment, so that they gradually expand their skills and techniques.

If your company is too small or has too few writers to make in-house training sessions cost-effective, you can consider using periodic outside consulting help. Many smaller companies make use of the continuing education conferences and seminars conducted by universities and technical institutes. Working with consultants and attending two- or three-day seminars gives writers a chance to learn and exchange ideas with other professionals, to bring themselves up to date on product liability, and to practice their writing skills. These comments, for example, came from writers who had been given outside help, either through consulting or at a conference:

"This is my father's company, and I got the manual-writing job. After attending this conference, I'm going to redo the whole thing. It scares me to see how many mistakes I've made. If someone got injured or killed with our equipment, we wouldn't stand a chance in court with our safety warnings. We're small. We could be wiped out with one lawsuit." (From a writer for a small, independent, industrial crane firm)

"I can't believe that I've been doing this job for over three years, and I never knew the range of choices I had for visuals. I didn't know that a manual could be used as legal evidence either. Why didn't somebody tell me?" (From a writer for a large medical equipment company)

If your company makes no provisions for organized training sessions, you can still ease the writing process a great deal by creating style handbooks, writer guidelines, and fact sheets listing steps involved in the manual process. Managers or editors can provide such books for their writers, and solo writers can create their own handbooks to systematize the procedures they plan to use. For example, one company's fact sheet, given to writers before work begins on the manual, contains the following information:

- Product name and number
- Deadline dates for completed manuscript in rough form
- Format specifications (column and page width, margins, type size, specifications for photos and drawings)
- Schedule and locations for viewing the product and for hands-on practice with mechanisms of the product
- Style guidelines (average sentence length, vocabulary and language level, use of active-voice verbs, etc.)
- Notification of what the other segments of the manual will be and which writer is assigned to those segments (especially important for cross-reference work or for machine systems that interact)
- List of phone numbers and names of people who can provide information
- References to materials on file that might be reused

Organizational Settings that Affect Writers

The setting and organizational structure in which a writer operates can be the single most important factor in good manual production. Structures affecting manual writers vary enormously from company to company. These variations are sometimes attributable to the size of the company, to managerial philosophy, or to the maturity of the product. Here are some of the patterns that affect the manual writer's job.

Large Companies

The very large company typically has a divisional organization, a diversity of products at scattered geographical locations, and separate cadres of technical writers specializing in manuals for each product category. Further, by the time a company goes national or international, its product line is usually "mature"; that is, the product has been around for some time, and the vocabulary for its parts and systems is quite well established and standardized. (Notable exceptions are the quick-growth electronic and computer technologies and those companies specializing in development of brand-new experimental products.)

Advantages   Very large companies with a team of technical writers whose sole responsibility is manual publication have the luxury of identifying and selecting good communicators from their own ranks. Or, when they choose to hire new employees, these companies usually have developed interviewing and testing systems to help them select the most qualified applicants. Quite often, in large companies, writers come up through the ranks, transferring from parts or service manual writing or from positions in product safety, marketing, or advertising. They bring to the job an in-depth knowledge of the product. Large companies are also able, through their service publication managers and editors, to identify writers who need help with their writing skills. That help is provided by one-on-one editorial assistance, on-the-job orientation, and periodic training sessions.

The publication capabilities of large companies often exceed those found in the formal publishing world. Fully equipped photographic labs; sophisticated printing machines; computerized systems for layout, format, and translation; full-color duplicating machines; and in-house personnel specializing in art and technical drawing, slide production, film, and videotape—all these are tools of the trade available at many large installations.

Disadvantages   There are also disadvantages to being a manual writer in a large company. Writers may have less autonomy and considerably less flexibility in deciding how best to do their job. If they are at widely scattered locations, they find that information takes longer to travel, filing systems may become nightmarishly complicated, and people and information may be harder to tap. If the large company is also decentralized, writing quality may be difficult to control. The manuals produced in Kentucky, for example, may be markedly different in quality and style from those produced in Florida. Because the large company tends to be more rigidly hierarchical, a decision to correct an error or to change the way manuals are done may take years, rather than months, to put into operation. In brief, what is gained through bigness, diversity, and sophistication may be lost through unwieldiness and lack of coordination.

Small and Intermediate-size Companies

The small and intermediate-size company typically has one or only a few locations. Such companies tend to be regional and centralized and to have a limited product line. Quite often the product is young and innovative, and consequently there may be no old manuals to use as guides, and no well-established vocabulary for parts and systems.

Advantages   For the writer, the small company can be an exciting and challenging place to work. A young product demands a fresh approach to the manual, and the writer can literally be the creator of the vocabulary and the approach. Further, writers are less likely to have to deal with inertia or with ''we've always done it this way'' frustration. Designers and engineers are likely to be more accessible to answer questions, and decision making is usually more fluid and flexible because the small company hierarchy has fewer layers. In fact, some of the most inventive ideas for manual production and layout come from the small companies lucky enough to have creative writers who had to build a first-time manual from the ground up.

Disadvantages   Many small and intermediate-size companies assign the manual writing to a single individual or to a small group of writers. These writers may be confronted with an awesome array of tasks. They must learn the technology of the product, plan layout, write text and safety messages, arrange for art work, photos, and drawings, negotiate with printers, edit, choose paper stock and type faces—and often do their own typing.

Publication support systems may be spotty in the small company (often little more than a typewriter, a desk top, and a corner in an office), and much of the production work must be contracted for. Manual writers who work ''solo'' feel the pressure of multiple responsibilities and are often rushed and isolated.

Managing the Work Place

Publication managers need to be alert to the importance of the work place, its structure, and its decision-making processes. It is possible to make vast improvements in writers' effectiveness with attention to such details as adequate work space and illumination, systems of information gathering, and acquisition of proper tools of the trade.

Establishing Lines of Authority

Every company has its own internal peculiarities, its hierarchies and pecking orders. Writers work within that pecking order, and situations will inevitably arise in which one person or unit has priority over another. Most writers can live comfortably with lines of authority, if they know what they are. What employees (writers included) find difficult are confused, pass-the-buck procedures in which the lines of authority are never articulated or clearly established.

As a supervisor, you may know where final authority lies for decisions on the manual, but you may neglect to convey that information to your writers. You should try to let your writers know about situations in which their decisions are likely to be superseded by someone with higher authority.

In manual writing, the most common problems with lines of authority arise in the following procedures:

- Determining who will have final say and sign-off on the manual's technical accuracy
- Deciding on appropriate language levels for manuals (engineers and lawyers are often disturbed by the simplicity required for general public users)
- Deciding on final authority when distinctions must be made between legal safeguards and engineering safeguards
- Deciding on final authority when an editor and a writer disagree sharply on word choice, format, or stylistic preference

Whenever possible, decisions like these should be made by discussion and consensus, with writers included in the discussion. But when negotiation is clearly not an option, let writers know where final authority lies.

What the Writer Needs

In Chapter 1, we discussed the writer's two basic needs: information and time. In this chapter, we added a third: training. If you are a manager of service publications, you should provide writers with as much assistance as possible. They should have

- Access to information
- Adequate time to do good work
- Training (either in-house or outside)

We suggest that you reread the suggestions on information gathering and scheduling in Chapter 1. Many of the techniques suggested there involve tasks for which you, as a supervisor, may bear chief responsibility.

Most of the technical writers we talk to are eager to do competent work. They are also quick to sense whether management seems to be "for" or "against" them. As the supervisor of manual production, you should be your writers' chief advocate in insisting on information access, training, and time.

Recognition for the Technical Writer

To the list of basic writers' needs, we must add a fourth: *recognition*.

Times are changing. For many companies, the manual used to be regarded as a bothersome necessity that got written at the last minute. Such attitudes were reflected in the scant time and money allotted to the manual and the meager recognition given to writers.

Industries are beginning to recognize that manual writers are the bridge builders between the product and the consumer. After your product has left the dealer's store, the manual becomes the interpreter of your product. Without the manual, the consumer must make a phone call or a trip back to the seller for help. As products grow more complex, as formerly simple mechanical devices are steadily being electronically controlled and computerized, and as manuals are interpreted by courts as significant legal evidence, the technical writer's work is becoming more valuable—and more highly valued. The new attitudes are reflected in better salaries, more investment in writing training programs, and better integration of the technical writer into the mainstream of company organizational structures.

The coalition of technical writers into a cohesive profession has also begun to take shape with conferences, seminars, professional societies, newsletters, and books. We hope that this book will prove to be a help to writers who have chosen technical writing as their profession.

Index

Analysis of the user (*see* User analysis)
Assessment of manual use, 23–30 (*see also* Feedback techniques)
Audience, 113, 117, 128 (*see also* Users)
Bilingual labeling, of toxic and dangerous products, 138 (*see also* Warnings)
Binding, 72, 131, 132

Callouts, 89
Charts:
    design principles, 54, 55
    in translated manuals, 140, 142, 143, 146, 147
    uses of, 53–55 (*see also* Wordiness, how to correct)
Checklists:
    of user characteristics and distinctions, 30, 31
    for safety warnings, 109, 110
    for user questions as organizers of the manual, 31
    for visuals, 91
    writers', 7, 9
Circuit diagrams, 125, 127, 128
Column length, 64
Column width, 62–64
Combining writing strategies, 47–53 (*see also* Syntax strategies; Writing strategies)
Company structures:
    centralized, 155, 156
    decentralized, 152, 154, 155
Comparison-contrast sentences, 43
Conditions of manual use, 19, 30 (*see also* Format; Binding)
Conference time, 5–7
Consistency:
    in terminology and nomenclature, 140
    of verb forms, 42, 43
Continuous-tone art, 76, (Fig. 5.3) 79
Cover, 72, 73, 131
Cross-referencing, 70–72
    in index, 131
    in translated manuals, 142, 143, 146, 147
    writer interaction, 153
Cultural differences, effect on manual use, 137
Cutaway diagrams, 77, (Fig. 5.1) 77, 82, 125, (Fig. 7.4) 127

Dangers, open and obvious, 95, 96
Deadlines, 5–8, 10–12, 153
Decision-making, 155–157
Deere and Company, 89, 90, 152

Description of mechanisms and processes, 46, 47 (*see also* Sequencing)
Developing nations, 136, 141, 142
Dictionaries, of industry terms, 139, 140
Documentation, of receipt of manual, 109
Drawings, 76–81
Duty to warn, 95, 96

Editing, 4, 10
Editors, 152, 157, 158 (*see also* Managers)
Engineering drawing, example, (Fig. 5.6) 82
Engineers, relationship of, to writers, 5, 7
Examples, as writing strategy, 57, (Fig. 3.9) 58
Exploded diagrams:
    divided into sections, 125, (Fig. 7.3) 126
    example, (Fig. 5.2) 78
    use of, 76

Fact sheets, for writers, 153
Feedback techniques:
    company troubleshooters, 23
    follow-up surveys, 23
    hidden camera, 23
    interviews with users, 26–30
    person-on-the-street, 23
Field modifications, 121, 122
Files, of manuals, 152, 153
Final copy, preparation of, 10
FMC Corporation, 98
Foreign languages, 134, 135 (*see also* Translated manuals)
Format:
    defined, 62
    in service manuals, 129–131
    standardized, 64
    in translated manuals, (Fig. 8.1) 142–143, 144
    two-column, 64, 144
Front matter, 108

General public products, categories of, 17
General public users:
    defined, 17, 18
    manuals for, 13
    sample of writing style for, (Ex. 2.2) 25
General-to-specific organization, 43, 44, 122
Generic manuals, problems of, 29, 30
Glossaries, for translated manuals, 139 (*see also* Standardized vocabulary)
Guidelines:
    for general public manuals, 20
    for professional manuals, 16

159